International Conference on Control Engineering and Artificial Intelligence (CCEAI 2018)

Journal of Physics: Conference Series Volume 976

Boracay, Philippines
19 – 21 January 2018

ISBN: 978-1-5108-5933-3
ISSN: 1742-6588

Printed from e-media with permission by:

Curran Associates, Inc.
57 Morehouse Lane
Red Hook, NY 12571

Some format issues inherent in the e-media version may also appear in this print version.

Printed by Curran Associates, Inc. (2018)

For permission requests, please contact the Institute of Physics at the address below.

Institute of Physics
Dirac House, Temple Back
Bristol BS1 6BE UK

Phone: 44 1 17 929 7481
Fax: 44 1 17 920 0979

techtracking@iop.org

Additional copies of this publication are available from:

Curran Associates, Inc.
57 Morehouse Lane
Red Hook, NY 12571 USA
Phone: 845-758-0400
Fax: 845-758-2633
Email: curran@proceedings.com
Web: www.proceedings.com

Table of contents

Volume 976

2018 International Conference on Control Engineering and Artificial Intelligence (CCEAI 2018)

19–21 January 2018, Boracay, Philippines

Accepted papers received: 19 February 2018
Published online: 2 March 2018

Preface

Papers

Control Engineering and Artificial Intelligence

IOP Conf. Series: Journal of Physics: Conf. Series **976** (2018) 011001 doi:10.1088/1742-6596/976/1/011001

PREFACE

CCEAI 2018 was held in Boracay, the Philippines, during 19-21 January 2018. CCEAI 2018 was organized by Asia Pacific Institute of Science and Engineering. The conference provides a useful and wide platform both for displaying the latest research and for exchanging research results and thoughts in Control Engineering and Artificial Intelligence. The participants of the conference were from almost every part of the world, with a background of either academia or industry, including well-known enterprises. The success and prosperity of the conference is reflected by the high level of the papers received.

The proceedings are a compilation of the accepted papers and represent interesting outcomes of research and applications of control engineering and artificial intelligence.

We would like to acknowledge all of those who supported CCEAI 2018. All the individual and institutional help was very important for the success of the conference. Especially we would like to thank the organizing committee for their valuable advice on the organization and effective peer review of the papers.

We hope that CCEAI will be a forum for excellent discussions that will put forward new ideas and promote collaborative research. We also hope that the proceedings will serve as an important source of references and knowledge, which will support advancement of science and engineering as well as development of new products and processes.

Prof. Yong Yue
Xi'an Jiaotong-Liverpool University, China

IOP Conf. Series: Journal of Physics: Conf. Series **976** (2018) 011002 doi:10.1088/1742-6596/976/1/011002

COMMITTEES

Conference Committee Co-Chairs

Desineni Subbaram Naidu, University of Minnesota, USA

Debao Zhou, University of Minnesota, USA

Program Committee Co-Chairs

Timothy Wolfe Bretl, University of Illinois at Urbana-Champaign, USA

Manuel Jr. Ramos, University of the Philippines Diliman, Philippines

Advisory Committee

Naira Hovakimyan, University of Illinois at Urbana-Champaign, USA

Michael R. M. Jenkin, York University, Canada

Publication Chair

Yong Yue Xi'an Jiaotong-Liverpool University, China

International Technical Committees

Samir Mekid, King Fahd University Petroleum & Minerals, Saudi Arabia

Syaimak Abdul Shukor , Universiti Kebangsaan Malaysia, Malaysia

Mahardhika Pratama , Nanyang Technological University, Singapore

Siti Norul Huda Sheikh Abdullah, Universiti Kebangsaan Malaysia, Malaysia

Erma Rahayu Binti Mohd Faizal Abdullah, University of Malaya, Malaysia

Woo Chaw Seng, University of Malaya, Malaysia

Siti Soraya Binti Abdul Rahman, University of Malaya, Malaysia

Aznul Qalid Bin Md Sabri, University of Malaya, Malaysia

Azizi Abdullah, Universiti Kebangsaan Malaysia,Malaysia

Maurizio Coan, University of Applied Sciences and Arts of Western Switzerland, Switzerland

Yukun Liu, Hebei University of Science and Technology, China

Yongqiang Gao, Ohlone College, USA

Zati Hakim Binti Azizul Hasan, University of Malaya, Malaysia

IOP Conf. Series: Journal of Physics: Conf. Series **976** (2018) 011003 doi:10.1088/1742-6596/976/1/011003

Peer review statement

All papers published in this volume of *Journal of Physics: Conference Series* have been peer reviewed through processes administered by the proceedings Editors. Reviews were conducted by expert referees to the professional and scientific standards expected of a proceedings journal published by IOP Publishing.

IOP Conf. Series: Journal of Physics: Conf. Series **976** (2018) 012001 doi:10.1088/1742-6596/976/1/012001

Investigation on the Nonlinear Control System of High-Pressure Common Rail (HPCR) System in a Diesel Engine

Le Cai, Xiaobing Mao and Zhexuan Ma

School of Energy and Power Engineering, Wuhan University of Technology, Wuhan 430063, China

E-mail address: maoxiaobing2009@163.com

Abstract. This study first constructed the nonlinear mathematical model of the high-pressure common rail (HPCR) system in the diesel engine. Then, the nonlinear state transformation was performed using the flow's calculation and the standard state space equation was acquired. Based on sliding-mode variable structure control (SMVSC) theory, a sliding-mode controller for nonlinear systems was designed for achieving the control of common rail pressure and the diesel engine's rotational speed. Finally, on the simulation platform of MATLAB, the designed nonlinear HPCR system was simulated. The simulation results demonstrate that sliding-mode variable structure control algorithm shows favorable control performances and overcome the shortcomings of traditional PID control in overshoot, parameter adjustment, system precision, adjustment time and ascending time.

1. Introduction

Bosch from Germany, Denso from Japan, Delphi from USA and Continental AG from Germany are four main global suppliers of common-rail electronically controlled fuel injection systems. Additionally, Caterpillar and Cummins from USA, and LIEBHERR and L'orange from Germany are also four manufacturers of common-rail electronically controlled fuel injection systems for self-use. Wärtsilä from Swiss and MAN DIESEL & TURBO from Germany launched low-speed marine diesel engines with electronically-controlled injection systems. Taking RT-Flex diesel engine manufactured by Wärtsilä Company as an example, 100MPa fuel in high-pressure common rail and 20MPa servo oil in medium-pressure common rail are transported directly by the plunger-type high-pressure fuel injection pump, which is driven by the transmission gear via a crankshaft; the pressure in two fuel rails can be adjusted by the electronic actuator and the connecting rod according to the preset pressure signal in Wätsilä Electric Control System. MAN DIESEL & TURBO Company uses medium-pressure common-rail (for servo oil) hydraulic-booster electronically controlled fuel injection system, in which the hydraulic unit provides the medium servo oil (i.e., lubricating oil) at 20MPa, and low-pressure fuel is boosted to 100MPa via a boosting piston and fuel injection is controlled by a FIVA valve.

Many control systems show nonlinear properties. Sometimes, the nonlinearity is inherent because of the object's dynamic properties. Although some objects themselves are linear, nonlinear laws are consciously introduced for achieving high-quality control. Strictly speaking, nonlinearity is ubiquitous and nonlinear systems are most general systems, while linear systems are only special cases. A common-rail electrically-controllable fuel injection system involves multiple parameters such as common-rail pressure, fuel-injection quantity and the rotational speed of the engine, which is undoubtedly a complex nonlinear system.

Variable-structure control is an advance control algorithm based on accurate mathematical model[1]. As a comprehensive method for nonlinear control systems, variable-structure control is practical and can acquire a series of favorable properties and qualities. Sliding mode variable structure control (SMVSC) is a control strategy for variable-structure control systems, which is also a comprehensive method in modern control theory based on phase plane method. The basic idea is that the system's motion first reaches a certain switching surface, then, the motion follows a sliding mode and gradually approaches the origin, and finally, the control goal can be realized when the sliding motion is endowed with good character.

At present, the main control method of diesel engine running parameters is PID[2]. This paper used Sliding mode variable structure control algorithm to study the nonlinear diesel engine.

2. Control of diesel engine common rail

HPCR system in a diesel engine is a hybrid linear system combining discrete events and continuous processes, which mainly consists of low-pressure fuel transfer pump, high-pressure fuel pump, high-pressure fuel pipe, fuel common-rail pipe, pressure regulator, filter, fuel injector and rail-pressure control valve (i.e., PCV value). At a high rotational speed, fuel was fed more frequently in unit time, and therefore, the common-rail system maintained a high common-rail fuel pressure. It means that the fuel pressure in a HPCR system mainly depends on the opening degree of the PCV value. Additionally, the opening degree can also regulate the fuel transfer quantity so as to achieve the regulation of common-rail pressure[3].

2.1. Establishment of the nonlinear model of high-press common rail(HPCR) system

HPCR oil pressure determines both fuel injection quantity and fuel return quantity. On the other hand, the opening degree of the PCV value depends on the pulse width and starting time of the PCV value in the electronic circuit. In fact, the fuel quantity that is transported to fuel rail by the fuel transfer pump depends on the correct timing of the current on the PCV valve. Therefore, we should clarify the nonlinear relationships among fuel pressure, fuel injection flow, fuel supply flow and fuel return flow. The common-rail pressure is subjected to the effects of fuel supply quantity of the high-pressure fuel pump, fuel return flow and fuel injection quantity of the injector.

Accordingly, for the convenience in the design of control law and the related stability analysis, the common-rail pressure model for a diesel engine can be simplified. The relationship among common-rail pressure, fuel injection flow, fuel supply flow and fuel return flow can be written as:

$$\frac{\mathrm{d}P_{cr}}{\mathrm{d}t} = \frac{K}{V}(Q_1 - Q_2 - Q_3) \tag{1}$$

where Q_1 denotes the fuel supply flow of a HVCR pump, Q_2 denotes the injector's fuel injection flow, Q_3 denotes fuel return flow of the PCV valve, P_{cr} denotes the fuel pressure in the HVCR pipe, K denotes the diesel's elastic modulus and V denotes the high-pressure common rail volume.

For accurately acquiring the quantitative relations among common rail pressure, the engine's rotational speed and fuel injection quantity, a 6L16/24 diesel was selected for the experiments on a HPCR testing platform. The fuel supply and return data at different common-rail pressures of 80MPa, 100MPa, 120MPa and 140MPa were recorded for further analysis.

Fuel supply flow mainly depends on the rotational speed of the high-pressure fuel pump which equals to half of the diesel's rotational speed n and the common rail pressure P_{cr} .Therefore, the following expression can be derived:

$$Q_1 = k_1 n + c_1 \tag{2}$$

Then, a quadratic polynomial of P_{cr} relative to k_1 and c_1 can be acquired through fitting, which can be written as:

$$k_1 = -0.0000000392 P_{cr}^{\ 2} + 0.00000273 P_{cr} + 0.001065 \tag{3}$$

$$c_1 = 0.0000581 P_{cr}^{\ 2} - 0.00356 P_{cr} - 0.072 \tag{4}$$

If the injector's fuel injection flow q is set as the circulating fuel injection flow, the following expression can be acquired:

$$Q_2 = q \tag{5}$$

The fuel return quantity increases with the rising of HPCR pressure. In this study, by neglecting the other factors of the equipment, fuel return quantity is directly determined by HPCR pressure:

$$Q_3 = 0.00416 P_{cr} \tag{6}$$

According to equations (1)-(6), the mathematical model of a HPCR nonlinear system can be described as:

$$\frac{\mathrm{d}P_{cr}}{\mathrm{d}t} = (a_1 P_{cr}^2 + a_2 P_{cr} + a_3) n + a_4 P_{cr}^2 + a_5 P_{cr} + a_6 + a_7 q \tag{7}$$

where a_1=—0.00392, a_2=0.273, a_3=106.5, a_4=5.81, a_5=—35.6, a_6=—720, a_7=929.6

Obviously, the differential equation not only includes the nonlinear quadratic component of HPCR pressure but also connects with the diesel's rotational speed and fuel injection quantity.

The differential equation of the rotational speed control system is also required. Similarly, using curve fitting method, the circulating fuel injection quantity data and average indicating pressure p_i at different rotational speeds (900 r/min, 1000 r/min, 1100 r/min and 1200 r/min, respectively) were analyzed. The average indicating pressure can be expressed as:

$$p_i = k_2 q + c_2 \tag{8}$$

Then, we conducted linear fitting on the coefficients and polynomial fitting on the constant, and the following expression can be acquired:

$$k_2 = 3.9 \tag{9}$$

$$c_2 = 0.00003346 n^2 - 0.107 n - 71.5294 \tag{10}$$

The variation of diesel rotating speed directly depends on the difference between output torque M and load torque M_l. According to the D' Alembert fundamental equation of motion[4], the following expression can be acquired:

$$J \frac{\mathrm{d}w}{\mathrm{d}t} = M - M_l \tag{11}$$

Assuming that the characteristic of load torque M_l is the vehicle's stable road characteristic, the following expression can be derived:

$$M_l = l \cdot n^2 \tag{12}$$

Let l= 0.00007 and $J = 5 kg \cdot m^2$, the nonlinear mathematic model for the diesel's rotational speed controlling can be written as:

$$\frac{\mathrm{d}n}{\mathrm{d}t} = b_1 n^2 + b_2 n + b_3 + b_4 n q \tag{13}$$

where b_1=—0.0001, b_2=—0.144, b_3=—114.7, b_4=4.2

According to equations (7) and (13), the mathematic model of a nonlinear HPCR control system can be written as:

$$\begin{cases} \frac{\mathrm{d}P_{cr}}{\mathrm{d}t} = (a_1 P_{cr}^2 + a_2 P_{cr} + a_3)n + a_4 P_{cr}^2 + a_5 P_{cr} + a_6 + a_7 q \\ \frac{\mathrm{d}n}{\mathrm{d}t} = b_1 n^2 + b_2 n + b_3 + b_4 q \end{cases} \tag{14}$$

This model is characterized by the quadratic component with nonlinearity and includes two state variables to be controlled, namely, common rail pressure and the diesel's rotational speed.

2.2. Design of a sliding mode controller (SMC)

For the system as described in equation (14), the state variable can be set as:

$$x_1 = \int_0^t e_1 \mathrm{d}t, \quad x_2 = e_1 = P_{cr0} - P_{cr},$$

where P_{cr0} denotes the preset common rail pressure and e_1 denotes the pressure difference.

$$x_3 = \int_0^t e_2 \mathrm{d}t, \quad x_4 = e_2 = n_0 - n,$$

where n_0 denotes the preset value of the diesel's rotational speed and e_2 denotes the difference of rotational speed.

Therefore, the state equation of the system can be rewritten as:

$$\begin{cases} \dot{x}_1 = x_2 \\ \dot{x}_2 = -(a_1 P_{cr}^2 + a_2 P_{cr} + a_3)n - (a_4 P_{cr}^2 + a_5 P_{cr} + a_6) - a_7 q \\ \dot{x}_3 = x_4 \\ \dot{x}_4 = -(b_1 n^2 + b_2 n + b_3) - b_4 q \end{cases} \tag{15}$$

In a standard form of state equation, the above expression can be rewritten as:

$$\begin{cases} \dot{x}_1 = x_2 \\ \dot{x}_2 = -d_1 x_2^2 x_4 + d_2 x_2^2 + d_3 x_2 x_4 + d_4 x_2 + d_5 x_4 + d_6 + d_7 u \\ \dot{x}_3 = x_4 \\ \dot{x}_4 = m_1 x_4^2 + m_2 x_4 + m_3 + m_4 u \end{cases} \tag{16}$$

where $d_1 = a_1$, $d_2 = -(a_1 n_0 + a_4)$, $d_3 = -(2a_1 P_{cr0} + a_2)$, $d_4 = 2a_1 P_{cr0} n_0 + a_2 n_0 + 2a_4 P_{cr0} + a_5$, $d_5 = a_1 P_{cr0}^2 + a_2 P_{cr0} + a_3$, $d_6 = -(a_1 P_{cr0}^2 n_0 + a_2 P_{cr0} n_0 + a_3 n_0 + a_4 P_{cr0}^2 + a_5 P_{cr0} + a_6)$, $d_7 = -a_7$, $m_1 = -b_1$, $m_2 = 2b_1 n_0 + b_2$, $m_3 = -(b_1 n_0^2 + b_2 n_0 + b_3)$, $m_4 = -b_4$

Next, the system input equation with m inputs was considered (in practical applications, a type of affine systems were considered).

$$\frac{\mathrm{d}}{\mathrm{d}t}\mathbf{x} = \mathbf{f}(\mathbf{x}) + \sum_{i=1}^{m} \mathbf{g}_i(\mathbf{x})\mathbf{u}_i \quad \mathbf{x} \in R^n \tag{17}$$

where $\mathbf{f}(\mathbf{x}) = \begin{bmatrix} x_2 \\ -d_1 x_2^2 x_4 + d_2 x_2^2 + d_3 x_2 x_4 \\ + d_4 x_2 + d_5 x_4 + d_6 \\ x_4 \\ m_1 x_4^2 + m_2 x_4 + m_3 \end{bmatrix}$, $\mathbf{g}(\mathbf{x}) = \begin{bmatrix} 0 \\ d_7 \\ 0 \\ m_4 \end{bmatrix}$

If the above expression can be rewritten as the following expression using a nonlinear state transformation $\mathbf{z} = \boldsymbol{\lambda}(\mathbf{x})$:

$$\begin{cases} \dfrac{\mathrm{d}}{\mathrm{d}t} z_i = p_i(z_1, \cdots, z_n) \quad (i = 1, \cdots, n-1) \\ \dfrac{\mathrm{d}}{\mathrm{d}t} z_n = p_n(z_1, \cdots, z_n) + q_n(z_1, \cdots, z_n) u \end{cases} \tag{18}$$

the analysis of the system will be simpler.

($n - m = 3 - 2 = 1$) base vectors were added to $\mathbf{G}(x)$, and a matrix $\mathbf{G}^*(\mathbf{x}) \in R^{n \times n}$ was formed.

$$\mathbf{G}^*(\mathbf{x}) = [\mathbf{g}_1 \quad \mathbf{g}_2 \quad \mathbf{g}_3 \quad \mathbf{g}_4] = \begin{bmatrix} 1 & 0 & 0 & 0 \\ 0 & 1 & 0 & d_7 \\ 0 & 0 & 1 & 0 \\ 0 & 0 & 0 & m_4 \end{bmatrix} \tag{19}$$

When $\mathbf{x} = \mathbf{x}^0 = (0, 0, 0)$, $\mathbf{G}^*(\mathbf{x}^0)$ is nonsingular. The flows, $\Phi_{z_1}^{g_1}$, $\Phi_{z_2}^{g_2}$, $\Phi_{z_3}^{g_3}$ and $\Phi_{z_4}^{g_4}$ were calculated using the method of flow calculation. The integral flow of the vector field, $\mathbf{g}_1(\mathbf{x})$, was first calculated.

$$\begin{cases} \dot{x}_1 = 1 \\ \dot{x}_2 = 0 \\ \dot{x}_3 = 0 \\ \dot{x}_4 = 0 \end{cases} \quad \begin{cases} x_1(t) = t + x_1^0 \\ x_2(t) = x_2^0 \\ x_3(t) = x_3^0 \\ x_4(t) = x_4^0 \end{cases} \quad \Phi_{z_1}^{\mathbf{g}_1}(\mathbf{x}) = \begin{bmatrix} z_1 + x_1 \\ x_2 \\ x_3 \\ x_4 \end{bmatrix}$$

Similarly, the integral flows of the vector fields $\mathbf{g}_2(\mathbf{x})$ and $\mathbf{g}_3(\mathbf{x})$ were calculated, and the following solutions were obtained:

$$\Phi_{z_2}^{\mathbf{g}_2}(\mathbf{x}) = \begin{bmatrix} x_1 \\ z_2 + x_2 \\ x_3 \\ x_4 \end{bmatrix} \quad \Phi_{z_3}^{\mathbf{g}_3}(\mathbf{x}) = \begin{bmatrix} x_1 \\ x_2 \\ z_3 + x_3 \\ x_4 \end{bmatrix} \quad \Phi_{z_4}^{\mathbf{g}_4}(\mathbf{x}) = \begin{bmatrix} x_1 \\ d_7 z_4 + x_2 \\ x_3 \\ m_4 z_4 + x_4 \end{bmatrix}$$

Then, we made the composition of flow mapping and acquired the transformation function of coordinates:

$$\mathbf{x} = \mathbf{F}(\mathbf{z}_1, \mathbf{z}_2, \mathbf{z}_3, \mathbf{z}_4) = \Phi_{z_1}^{\mathbf{g}_1} \circ \Phi_{z_2}^{\mathbf{g}_2} \circ \Phi_{z_3}^{\mathbf{g}_3} \circ \Phi_{z_4}^{\mathbf{g}_4}(\mathbf{x}^0) \tag{20}$$

$$\begin{bmatrix} x_1 \\ x_2 \\ x_3 \\ x_4 \end{bmatrix} = \Phi_{z_1}^{\mathbf{g}_1} \circ \Phi_{z_2}^{\mathbf{g}_2} \circ \Phi_{z_3}^{\mathbf{g}_3} \circ \Phi_{z_4}^{\mathbf{g}_4} = \begin{bmatrix} z_1 + x_1^0 \\ d_7 z_4 + z_2 + x_2^0 \\ z_3 + x_3^0 \\ m_4 z_4 + x_4^0 \end{bmatrix} \tag{21}$$

Set $\mathbf{x}^0 = (0, 0, 0)$, the following expression can be acquired:

$$\begin{cases} x_1 = z_1 \\ x_2 = d_7 z_4 + z_2 \\ x_3 = z_3 \\ x_4 = m_4 z_4 \end{cases} \tag{22}$$

The inverse transformation can be written as:

$$\begin{cases} z_1 = x_1 \\ z_2 = x_2 - d_7 x_4 / m_4 \\ z_3 = x_3 \\ z_4 = x_4 / m_4 \end{cases} \tag{23}$$

in which it was assumed that

$$\boldsymbol{\lambda}(\mathbf{x}) = \begin{bmatrix} \lambda_1(\mathbf{x}) \\ \lambda_2(\mathbf{x}) \\ \lambda_3(\mathbf{x}) \\ \lambda_3(\mathbf{x}) \end{bmatrix} = \begin{bmatrix} x_1 \\ x_2 - \dfrac{d_7 x_4}{m_4} \\ x_3 \\ \dfrac{x_4}{m_4} \end{bmatrix} \tag{24}$$

$$\frac{\mathrm{d}}{\mathrm{d}t}\mathbf{z} = \frac{\partial}{\partial \mathbf{x}}\lambda \frac{\mathrm{d}}{\mathrm{d}t}\mathbf{x} = \frac{\partial \lambda}{\partial \mathbf{x}} f(\mathbf{x}) + \frac{\partial \lambda}{\partial \mathbf{x}} g(\mathbf{x})u = \begin{bmatrix} x_2 \\ -d_1 x_2^2 x_4 + d_2 x_2^2 + d_3 x_2 x_4 + d_4 x_2 + d_5 x_4 \\ + d_6 - \dfrac{d_7}{m_4}(m_1 x_4^2 + m_2 x_4 + m_3) \\ x_4 \\ \dfrac{1}{m_4}(m_1 x_4^2 + m_2 x_4 + m_3) \end{bmatrix} + \begin{bmatrix} 0 \\ 0 \\ 0 \\ 1 \end{bmatrix} u = \begin{bmatrix} p_1(\mathbf{z}) \\ p_2(\mathbf{z}) \\ p_3(\mathbf{z}) \\ p_4(\mathbf{z}) \end{bmatrix} + q(\mathbf{z})u \tag{25}$$

Using the linear switch function, the above expression can be rewritten as:

$$s(\mathbf{z}) = c_1 z_1 + c_2 z_2 + c_3 z_3 + z_4 \tag{26}$$

$$\frac{\mathrm{d}}{\mathrm{d}t} s = c_1 p_1(\mathbf{z}) + c_2 p_2(\mathbf{z}) + c_3 p_3(\mathbf{z}) + p_4(\mathbf{z}) + q(\mathbf{z})u \tag{27}$$

Based on the generalized sliding mode conditions, the sliding mode control can be described as:

IOP Conf. Series: Journal of Physics: Conf. Series **976** (2018) 012001 doi:10.1088/1742-6596/976/1/012001

$$u=\begin{cases} u^{+}(z)<-q^{-1}(z)[c_1p_1(\mathbf{z})+c_2p_2(\mathbf{z}) \\ \qquad +c_3p_3(\mathbf{z})+p_4(\mathbf{z})] \quad \text{when } s>0 \\ u^{-}(z)>-q^{-1}(z)[c_1p_1(\mathbf{z})+c_2p_2(\mathbf{z}) \\ \qquad +c_3p_3(\mathbf{z})+p_4(\mathbf{z})] \quad \text{when } s<0 \end{cases} \tag{28}$$

3. Software simulations

Next, using the CAGE toolbox in MATLAB, the original performance data were interpolated for solving the engine's performance data at the other operating points and determining the MAP chart of the simulation model. These MAP were then deposited in advance in the engine operating module on simulation platform.

In the first case, the diesel engine's rotational speed and fuel injection quantity were used as the input variables while the average indicating pressure was used as the output response. In the second case, the diesel engine's HPCR pressure and fuel injection quantity were used as the input variables while the fuel injection pulse width, the opening degree of PCV valve and the actual common rail pressure served as the output response. The diesel engine's HPCR pressure, rotational speed and fuel injection quantity data were then input to MAP charts for acquiring the calibrated target values.

The simulation program of the system is shown in Figure 1. Simulation results of the common rail pressure and speed of the diesel engine are shown in Figures 2-7.

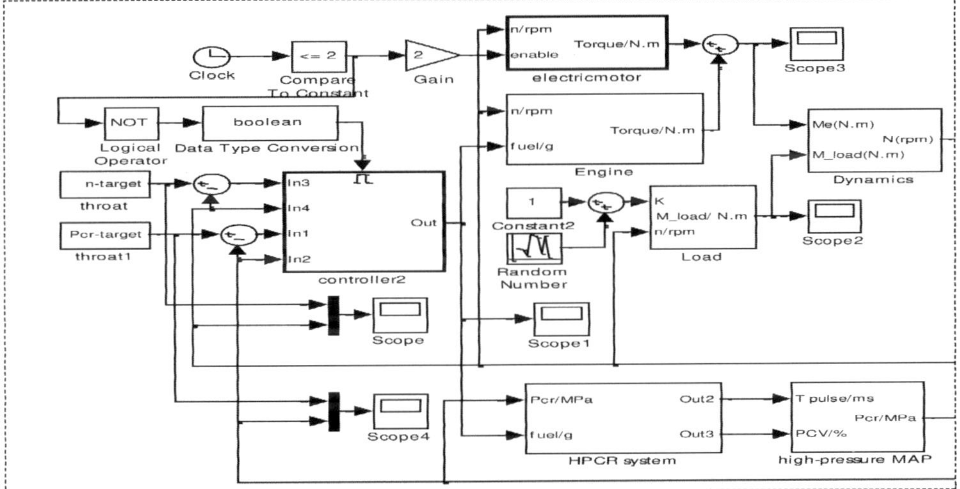

Figure 1. Simulation structure figure of the common rail and speed system.

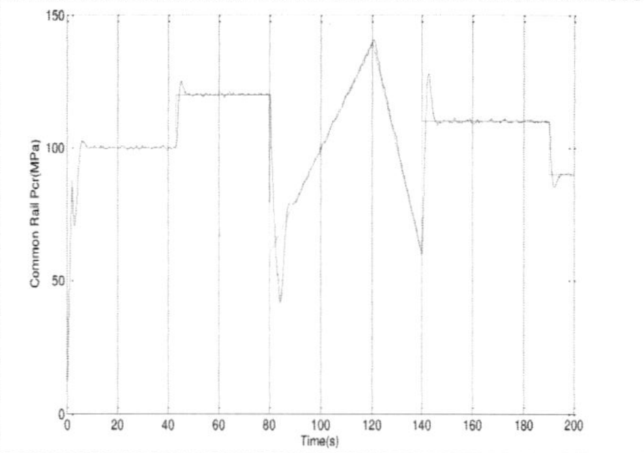

Figure 2. Common rail pressure simulation of PID.

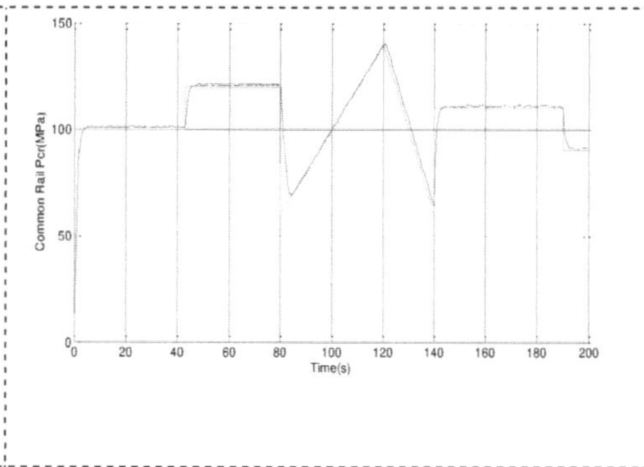

Figure 3. Common rail pressure simulation of SMVSC.

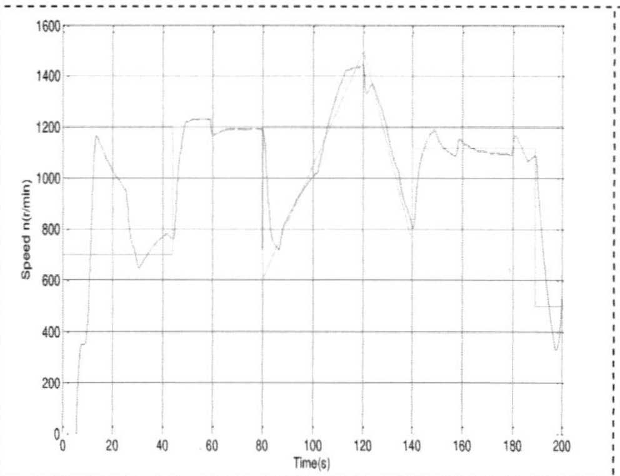

Figure 4. Rotational speed simulation of PID.

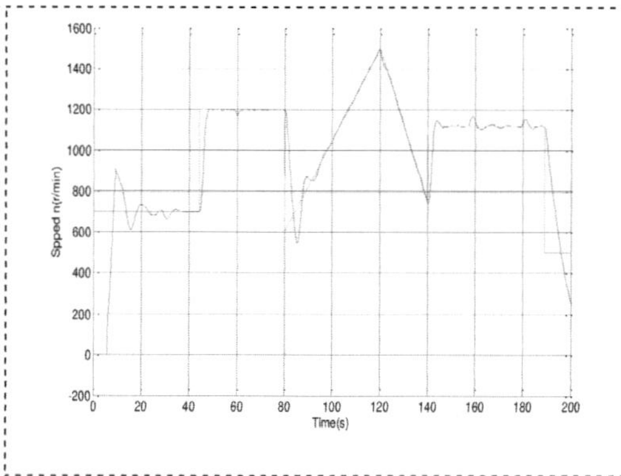

Figure 5. Rotational speed simulation of SMVSC.

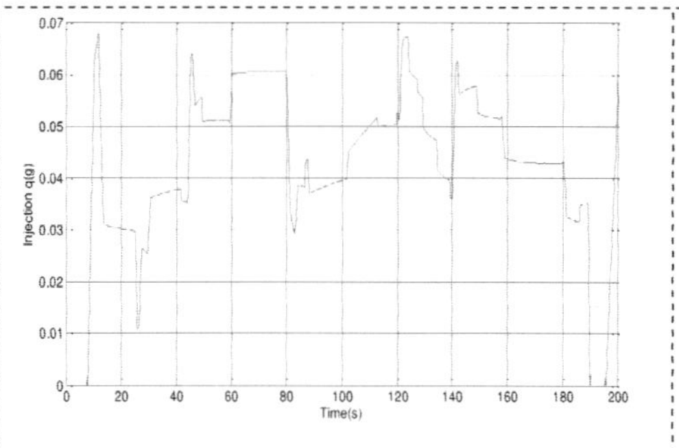

Figure 6. Variation of oil injection with PID.

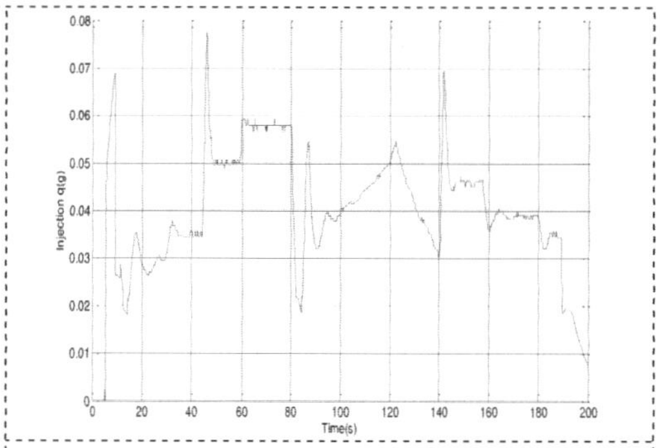

Figure 7. Variation of oil injection with SMVSC.

According to the simulation results, For SMVSC, the rotational speed overshoot was much smaller than PID control, especially in the initial stage of diesel engine starting, and speed fluctuated within 10 r/min in the stable stage. Additionally, the diesel engine's transient speed-regulating rate was smaller than 10%, the stable speed-regulation rate was about 0%, and the fluctuation ratio of rotational speed

IOP Conf. Series: Journal of Physics: Conf. Series **976** (2018) 012001 doi:10.1088/1742-6596/976/1/012001

was smaller than 1%. Under a sudden load-up of 100%, the stabilization time of 5% error was smaller than 4 seconds.

For the common rail pressure control, SMVSC varies within a range of 5%. The simulation results fully proved the feasibility of variable structure control theory in investigating the common rail and rotational speed systems of nonlinear diesel engine. Using variable structure control method, the system's nonlinear characteristics were well maintained.

4. Conclusions

In previous studies, the diesel engine's rotational speed, common rail pressure, fuel viscosity temperature and cooling water temperature were generally controlled by traditional PID strategy [5]. Although the control structure was simple, the control precision, stability, robustness and control performances were far from ideal.

The fundamental difference between sliding-mode variable structural control and conventional control lies in the control's discontinuity, i.e., a kind of switching characteristics that can make the system structure change at any time. Using sliding model variable structure control method, system shows a series of advantages including fast response, small overshoot and strong robustness.

This research project has been supported by the independent innovation research fund project (2017-ND-A1-03) of Wuhan University of Technology. I'd like to express my gratitude.

5. References

[1] Farouk N and Leghmizi 2012 Speed control system on marine diesel engine based on a self-tuning fuzzy PID controller *Research Journal of Applied Sciences, Engineering and Technology* vol 4 chapter 6 pp 686-690

[2] Xu F and Chen H 2015 Fast Nonlinear model predictive control on FPGA using particle swarm optimization *IEEE Transactions on Industrial Electronics* vol 63 chapter 5 pp 310-321

[3] Fakhari V and Ohadi A 2013 A robust adaptive control scheme for an active mount using a dynamic engine model *Journal of Vibration & Control* vol 21 chapter 11 pp 1-23

[4] Sun Z and Kuo T W 2010 Transient control of electro-hydraulic fully flexible engine valve actuation system *IEEE Transactions on Control Systems Technology* vol 18 chapter 3 pp 613-621

[5] Wang Y and Chen B 2011 Simulation research on intelligent control for propulsion system of large ship *Oceans IEEE* vol 5 chapter 7 pp 1-7

Monocular Visual Odometry Based on Trifocal Tensor Constraint

Y J Chen, G L Yang, Y X Jiang and X Y Liu

Science and Technology on Inertial Lab, Beihang University, 37Xueyuan Road, Beijing, China

cyj_tju@163.com

Abstract. For the problem of real-time precise localization in the urban street, a monocular visual odometry based on Extend Kalman fusion of optical-flow tracking and trifocal tensor constraint is proposed. To diminish the influence of moving object, such as pedestrian, we estimate the motion of the camera by extracting the features on the ground, which improves the robustness of the system. The observation equation based on trifocal tensor constraint is derived, which can form the Kalman filter alone with the state transition equation. An Extend Kalman filter is employed to cope with the nonlinear system. Experimental results demonstrate that, compares with Yu's 2-step EKF method, the algorithm is more accurate which meets the needs of real-time accurate localization in cities.

1. Introduction

Visual odometry (VO) and real-time monocular Simultaneous Localization and Mapping (SLAM) have become increasingly popular research topic [1]. They estimates the motion of the object relying on the input information of the camera. With the further development of image processing techniques, these algorithms have been gradually applied to robotics, unmanned aerial vehicles, and virtual reality applications [1].

Visual odometry can roughly be divided into monocular odometry and stereo odometry, or into feature-based method and direct method according to the algorithm [2]. Nister et al. [3] who proposed sparse methods for estimating the small movement of monocular by sequential frame-to-frame matching. More recently, Forster et al. [4] proposed semi-direct visual odometry (SVO), a real-time capable, direct method that be applied to micro-aerial-vehicle. They deducted the Gaussian mixture distribution model of the depth and reconstruct the 3D scene. The most recent Kalman filter based methods are by Yu et al [5, 6], the full covariance extended Kalman filter (EKF) is decoupled such that the computation efficiency is increased as a tradeoff in accuracy.

In this paper, we proposed a monocular visual odometry based on trifocal tensor constraint [7] which is real-time capable on CPUs. Our method tracks the interest points with Lucas-Kanade optical-flow [8] and form Kalman filter model using trifocal tensor constraint. It concurrently estimates the rigid-body motion through Extend Kalman filter. The proposed approach can improve the precision of the optical-flow and enhance system robust.

2. Preliminaries

2.1. Camera model

One point $P_w = (x_w \quad y_w \quad z_w)^T \in S$ in the world coordinates $S \in R^3$ map to the image coordinates $P_{uv} = (u \quad v)^T \in \Pi$ through the projection model as follows, where $\Pi \in R^2$ is the image domain.

$$\tilde{P}_{uv} = P\tilde{P}_w = K[R \quad t]\tilde{P}_w \tag{1}$$

Where $\tilde{P}_{uv}$ is in homogeneous coordinates and K denotes the camera intrinsic parameters which can be gained from calibration. The motion of the camera is expressed with its rotation matrix R and translation vector ***t***. P denotes the camera matrix.

2.2. Trifocal tensor constraint

The trifocal tensor to the three views is similar to the fundamental matrix to the two views. It describes the imaging geometric relationship which is independent of scene structure among three views. The constraint allows us to estimate the motion of the camera without recovering the 3D points from the world coordinates which can avoid the error from triangularization.

For the sake of simplicity, we assume that we get consecutive three images from one camera. A point ***X*** from world coordinates map to the three images X_A、X_B、X_C and the corresponding transformation matrix are denoted as T_A, T_B, T_C, respectively. And the camera matrixes of the three views are as follows:

$$P_A = KT_A \tag{2}$$

$$P_B = KT_B \tag{3}$$

$$P_C = KT_C \tag{4}$$

And the corresponding trifocal tensors are indicated below.

$$T_i^{qk} = (-1)^{i+1} \bullet \det \begin{pmatrix} \sim a^i \\ b^q \\ c^k \end{pmatrix} \tag{5}$$

Where $\sim a^i$ denotes P_A without i^{th} row, b^q denotes q^{th} row of P_B, c^k denotes k^{th} row of P_C.

After calculating the tensor matrix, we can track the corresponding feature points in the third image through the matched feature points from first and second image. As shown in figure 1, we construct a line $l_{B,j}$ going through a feature point X_B in second image and project this line on the scene surface to form a plane which construct a planar scene between first and second image. So we can track the X_C by estimating a homography.

$$X_C^k = X_A^k \cdot l_{B,j} \cdot T_i^{jk} \tag{6}$$

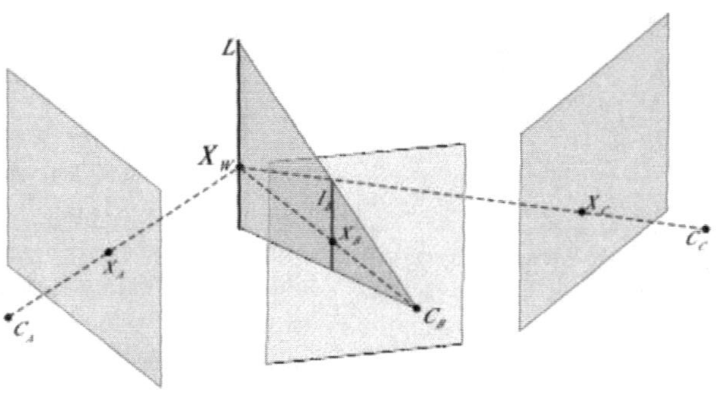

Figure 1. The feature correspondences in three views construct the trifocal tensor constraint.

3. Monocular visual odometry based on trifocal tensor constraint

The algorithm consist of three major modules: initialization, trifocal tensor tracking and motion estimation based on EKF as visualized in figure 2.

3.1. Initialization

We define the notation as follows. The image collected at timestep t is denoted with$\boldsymbol{f_t}$, the rate of frame isΔt. The feature correspondences of $\boldsymbol{f_1}$, $\boldsymbol{f_2}$ and $\boldsymbol{f_t}$ is denoted as $x_i, x_i^{'}, x_i^{''} (\mathrm{i}=1,2,...,\mathrm{N})$.

Because of scale uncertainty, monocular visual odometry need initialization and to be given sufficient translational camera movement between the first and second frame [1, 9]. In practice, we estimate the motion using the camera pointing to the ground, which can calculate the initial pose by homography. Based on plane hypothesis, we can recognize the points on the ground by calculating the primary homography and measure the inliers by RANSAC [10].

Firstly, we extract the FAST corner features from frame f_1 and track the feature correspondences by L-K optical flow in the next frame. As shown in figure 3, the pedestrians and cars are general at the top of the images since ground account for the main part of the images.

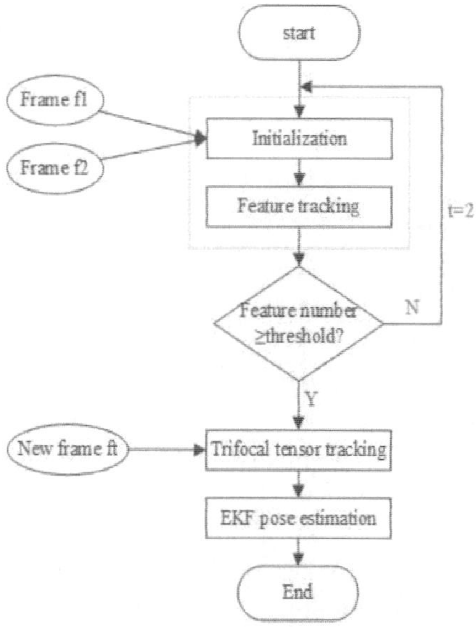

Figure 2. This is a figure providing an overview of the whole algorithm. The algorithm can be divided into three module: initialization, trifocal tensor tracking and motion estimation based on EKF.

Figure 3. This figure shows the result of the optical flow, which illustrates the features extracted from the ground are Detection of optical flow points on ground.

Based on these assumptions, we choose the four groups of correspondences at the bottom of the image as inliers and calculate the homography between these correspondences which denotes as H. Then we calculate the re-projection error for the rest of the points selected at random.

$$E(\mathrm{x}_i, \mathrm{x}_i^{'}, \mathrm{H}) = \left\| \mathrm{x}_i^{'} - H\,\mathrm{x}_i \right\| \tag{7}$$

The point is considered as outlier if the error exceeds the threshold we set before and deleted, which means that all the points used to estimate the motion of the camera are inliers. The initial guess of the f_2's pose is required through homography matrix decomposition and refined in the process of EKF.

3.2. Trifocal tensor tracking

After tracking the current frame, the frame f_1, f_2 and current frame compose the three views which are used to construct trifocal tensor constraint. The detailed trifocal tensor constraint can refer to [9]. The number of the feature correspondences tracked may decrease with the angle view change, so we need to restart EKF when the number of the features is under a minimum size in current frame. In practice, the process followed is that the tracking algorithm will restart with frames f_{t-1} and f_t. We will recalculate the feature correspondences. And then we reset t to 2. The new EKF process is then activated by frame f_{t+1}. The roll-back method keep the same scale among the whole algorithm.

3.3. Motion estimation based on EKF

The model parameters to be estimated are given:

$$y_k = \begin{bmatrix} t_{xk} & \dot{t}_{xk} & t_{yk} & \dot{t}_{yk} & t_{zk} & \dot{t}_{zk} & \alpha_k & \dot{\alpha}_k & \beta_k & \dot{\beta}_k & \gamma_k & \dot{\gamma}_k & t_{x2} & t_{y2} & t_{z2} & \alpha_2 & \beta_2 & \gamma_2 \end{bmatrix} \quad (8)$$

Where t_{xk}, t_{yk}, t_{zk} are the displacement alone x, y, z axis relative to the world coordinates, respectively. $\dot{t}_{xk}, \dot{t}_{yk}, \dot{t}_{zk}$ are corresponding velocity. $\alpha_k, \beta_k, \gamma_k$ are yaw, pitch and roll of the current frame, respectively. And $\dot{\alpha}_k, \dot{\beta}_k, \dot{\gamma}_k$ describe corresponding angular velocity.

The state transition equation and measurement equation are as follows:

$$y_{k+1} = f(\mathrm{y}_k) + \mathrm{w}_k \quad (9)$$

$$z_{k+1} = h(\mathrm{y}_{k+1}) + \mathrm{v}_{k+1} \quad (10)$$

The motion of the camera is complex, resulting in that f(·)is nonlinear state transition function. h(·) is nonlinear measurement function constructed by trifocal tensor constraint.

$z_{k+1} = \begin{bmatrix} u'_{1,t} & v'_{1,t} & u'_{2,t} & v'_{2,t} & \cdots & u'_{N,t} & v'_{N,t} \end{bmatrix}^T$ denotes 2N-dimensional measurement vector, which describes feature points in the current frame. $w_k \sim N(0, \mathrm{Q}_k)$ denotes system error, $v_{k+1} \sim N(0, \mathrm{R}_{k+1})$ denotes observation noise and N is the number of the feature correspondences.

To solve this model, the equation should linearization. In general, the motion of the camera is regarded as uniform motion with same speed and angular velocity based on the assumption of the high rate of frame. The initial guess of frame f2's pose is close to the true value. So the state equation can be given:

$$y_{k+1} = Ay_k + w_k \quad (11)$$

$$A = \mathrm{diag}\left\{ \begin{bmatrix} 1 & \Delta t \\ 0 & 1 \end{bmatrix} \begin{bmatrix} 1 & \Delta t \\ 0 & 1 \end{bmatrix} \begin{bmatrix} 1 & \Delta t \\ 0 & 1 \end{bmatrix} \begin{bmatrix} 1 & \Delta t \\ 0 & 1 \end{bmatrix} \begin{bmatrix} 1 & \Delta t \\ 0 & 1 \end{bmatrix} \begin{bmatrix} 1 & \Delta t \\ 0 & 1 \end{bmatrix} \begin{bmatrix} 1 & 0 \\ 0 & 1 \end{bmatrix} \begin{bmatrix} 1 & 0 \\ 0 & 1 \end{bmatrix} \begin{bmatrix} 1 & 0 \\ 0 & 1 \end{bmatrix} \right\} \quad (12)$$

The measurement equation is general linearized by Taylor expansion and calculating the Jacobian matrix, which is called extend Kalman filter.

For the measurement equation$h(y_{k+1})$:

$$x''_i = h(\mathrm{T}_k, \mathrm{l}'_2, \mathrm{x}_i) \quad (13)$$

According to the above analysis, the state variable updating procedure is shown as follows:

$$\hat{w}_{k+1,k} = A\hat{w}_{k,k} \quad (14)$$

$$P_{k+1,k} = AP_{k,k}A^T + Q_k \quad (15)$$

IOP Conf. Series: Journal of Physics: Conf. Series **976** (2018) 012002 doi:10.1088/1742-6596/976/1/012002

$$\hat{w}_{k+1,k+1} = \hat{w}_{k+1,k} + K(z_{k+1} - h(\hat{w}_{k+1,k})) \tag{16}$$

$$P_{k+1,k+1} = P_{k+1,k} - K\nabla J_y P_{k+1,k} \tag{17}$$

$$K = P_{k+1,k}\nabla J_y^T(\nabla J_y P_{k+1,k}\nabla J_y^T + R'_{k+1})^{-1} \tag{18}$$

Where ∇J_y is the Jacobian matrix of measurement equation at $\hat{w}_{k+1,k}$.

4. Experiment and result

Experiments were performed on videos recorded from a downward-looking camera attached to an electrombile in the school. Differential GPS is used to provide the ground-truth with 10Hz sampling frequency and 0.01m location error. The detailed parameters configuration is shown in table 1.

Table 1. Experiment configuration.

Camera	Image resolution	Frame-rate
bumblebee2	640×480	20Hz

4.1. Accuracy

We evaluate the accuracy on a video mentioned before and the result is illustrated in figure 4. The ground-truth for the trajectory originates from a differential GPS. The trajectory is 350 meters long and the car drives on average 7.56km/h. We contrast our approach with Yu's 2-step EKF algorithm and generate the plots. Figure 5 illustrates the position error over frame.

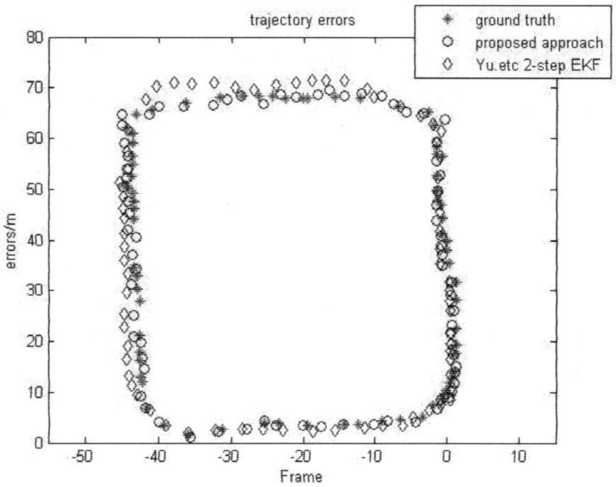

Figure 4. Comparison against the groundtruth.

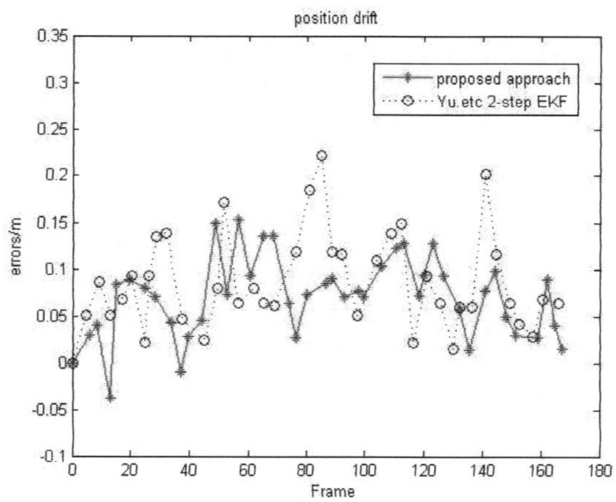

Figure 5. Position drift of the proposed method and comparison against Yu's method.

Overall, the proposed approach is more accurate than Yu's approach. We suspect the main reason is that our method combine the Kalman filter with trifocal tensor constraint without calculating the scene structure which eliminate the influence of triangularization and improve the system robustness. And in real scene, recognizing the ground reduce the influence of moving object, which increase the accuracy as well.

4.2. Running time

Table 2 show the running time required to operate the algorithm on the CPUs. Similarly, we compare the performance of proposed algorithm with Yu's 2-step EKF. From the table, we can see Yu's method is faster than our method. The main reason for that is that trifocal tensor constraint need to run at every frame.

Table 2. Running time comparison.

	Proposed approach	Yu's 2-step EKF
Running time (in second)	1.08	0.52

5. Conclusion

A monocular visual odometry based on trifocal tensor constraint has been proposed in this paper. It has a great improvement on the accuracy due to the combination of EKF and trifocal tensor constraint. The gain in accuracy is due to the fact that the scene structure doesn't need to be calculate which eliminate the influence of triangularization and improve the system robustness. The algorithm can be used in real-time precise localization in the urban street as it is capable of high frame-rate motion estimation and provide robust output combined with RANSAC.

Acknowledgments

The work described in this paper was supported by a grant from Science and Technology on Inertial Lab in Beihang University.

References

[1] Engel, J., Schöps, T., & Cremers, D. (2014). Lsd-slam: large-scale direct monocular slam. , *8690*, 834-849.

[2] Nister, D., Naroditsky, O., & Bergen, J. (2004). Visual odometry. *Computer Vision and Pattern Recognition, 2004. CVPR 2004. Proceedings of the 2004 IEEE Computer Society Conference on* (Vol.1, pp.I-652-I-659 Vol.1). IEEE.

[3] Royer, E., Lhuillier, M., Dhome, M., & Lavest, J. M. (2007). Monocular vision for mobile robot localization and autonomous navigation. *International Journal of Computer Vision*, 74(3), 237-260.

[4] Forster, C., Pizzoli, M., & Scaramuzza, D. (2014). SVO: Fast semi-direct monocular visual odometry. *IEEE International Conference on Robotics and Automation* (pp.15-22). IEEE.

[5] Yu, Y. K., Wong, K. H., & Chang, M. Y. (2008). A fast and robust simultaneous pose tracking and structure recovery algorithm for augmented reality applications. *International Conference on Image Processing* (Vol.2, pp.1029-1032 Vol.2). IEEE.

[6] Yu, Y. K., Wong, K. H., Ming, M., Chang, Y., & Or, S. H. (2006). Computing pose sequences directly from videos. 2773-2776.

[7] Hartley, R., & Zisserman, A. (2000). Multiple view geometry in computer vision. (Vol.30, pp.1865 - 1872). Cambridge University Press.

[8] Lucas, B. D., & Kanade, T. (1981). An iterative image registration technique with an application to stereo vision. *International Joint Conference on Artificial Intelligence* (Vol.73, pp.674-679). Morgan Kaufmann Publishers Inc.

[9] Scaramuzza, D., & Fraundorfer, F. (2011). Visual odometry: part i: the first 30 years and fundamentals. *IEEE Robotics & Automation Magazine*.

[10] Watanabe, T., Kamai, T., & Ishimaru, T. (2015). Robust estimation for camera homography by fuzzy RANSAC algorithm with reinforcement learning. *International Symposium on Soft Computing and Intelligent Systems* (Vol.19, pp.712-717). IEEE.

IOP Conf. Series: Journal of Physics: Conf. Series **976** (2018) 012003 doi:10.1088/1742-6596/976/1/012003

Rule-Based Multiple Target Monitoring in Robot Vision Using Partitioning

Davood Pour Yousefian Barfeh[*,1,2,3], Terrence A Lim[1], Editha C Jose[2], Felino P Lansigan[2], Eduardo R Mendoza[4] and Vladimir Y Mariano[5]

[1]Lyceum of the Philippines University-Laguna, Calamba City, Philippines
[2]University of the Philippines-Los Bosños, Laguna, 4031, Philippines
[3]Adamson University, 1000 Manila, Philippines
[4]Ludwig Maximilian University, 80539 Munich, Germany
[5]RMIT University, 702 Nguyen Van Linh, HCMC Vietnam

[*]davoodpb@gmail.com

Abstract. In this study, the researchers propose an algorithm and implement it as a robot-camera which not only detects many moving objects and captures them, but also follows the moving objects to continue capturing and monitoring. The robot uses Gaussian filtering to detect all motions and then applies partitioning on the scene to obtain location of all moving objects in three major areas and different sub-areas. Based on the partitioning and position of the moving objects, the robot might be in any of unsafe state, safe state, and over-safe state. Regarding each state, the robot performs proper movement following two cardinal rules to pursue monitoring the moving objects.

1. Introduction

There are three levels of algorithms in computer vision and digital image processing; low level vision algorithms, intermediate level vision algorithms, and high level vision algorithms [1–6]. As the names of levels show, the algorithms in high level are more difficult and more important, since these algorithms deal with perception and the context. Indeed, in these algorithms, the input and output of systems are symbolic representation of images' features [1–6]. The best instances of these algorithms can be found in the fields of target (moving object) tracking and motion detection in which the aim is to detect motion and moving object as target or to find the track of the target [1][6]. Beside importance and significance, there are several complications in target tracking or motion detection such as; occlusion, changes in illumination, change in background, clutter, occurrence of shadow and noisy video [7]. Regardless of these complications and challenges, motion detection and target tracking are interesting topics in computer vision and digital image processing, since they are bases in numerous fields such as supplying advanced interfaces between humans or between humans and devices, medicine, sports, military, and especially PTZ (Pan-Tilt-Zoom) or smart cameras [7–12].

2. Review of related studies and products

Numerous studies and researches are performed in the field of motion detection, target tracking and capturing. Moreover, there are several products in industry regarding these fields. All of these studies and products can be classified into four categories.

Category-A consists of studies and researches which are not real time, but systems and algorithms try to detect moving objects and then track the targets in recorded videos [13–19].

Category-B consists of studies and researches that work real time. Systems in these studies detect and track many targets using stationary cameras [20–22].

Category-C consists of studies performed based on a moving camera. Here, the systems focus on detecting and tracking of targets while the camera has a random movement [23–25]. The movement of camera is not dependent on motion of targets. In studies in Category-C same as the studies in Category-A and Category-B, the term "track" means that path of moving objects.

Category-D consists of industrial works and products in which the algorithms and systems use non-stationary cameras to detect and track moving object [26–28]. Systems in these products not only detect moving object, but also track, monitor and follow the moving object either by focal zooming or by movement. Products and systems of this category are able to detect, track, monitor and follow only one moving object. In Category-D, the term "track" means that following and monitoring of the moving object.

3. Statement of the problem

Most of the time, when it is talked about motion, there are many targets in the scene. In other words, there are many objects which are moving – not only one object. Therefore, we need a multiple moving object monitoring system that can capture and monitor many targets.

After considering all categories of researches, studies and products, the researchers realized that there is no algorithm or system to monitor and follow many targets. In fact, regarding monitoring and following target, the best studies are placed in Category-D in which systems are able to monitor and follow only one target, not more.

4. Proposed system

To solve the problem mentioned, the researchers propose a rule-based multiple target monitoring (RBMTM) system which is a robot system having a camera. RBMTM not only detects multiple moving object, but also moves forward, backward, left and right to continue capturing and monitoring of the moving objects.

In this study, researchers particularly design and implement an algorithm as a simple robot that;

a) Performs multiple moving object detection (motions detection), and

b) Tries to maintain monitoring of the moving objects using proper movements to right, left, forward and backward.

4.1. Motions detection

There are several methods and algorithms for motion detection. All of these methods work based on the comparison of the current video frame with one (or more) frame from the previous frame (or frames) [29].

The system RBMTM uses Gaussian filtering for motions detection to benefit simplicity and rapidity [29–32]. In digital image processing, Gaussian filtering (Gaussian smoothing) is the outcome of blurring a digital image using a Gaussian function that is extensively used in graphics tools to alleviate noise of image. Moreover, Gaussian function is used to reduce some details in image. In any case, the effect of the Gaussian function is generating a smooth blur resembling image [29–33].

4.2. Movement

After capturing the scene as a stream of images and detecting motions, the system RBMTM may need to move. The movement is the effort of the system to continue capturing and monitoring of moving objects. In fact, the system tries to chase all of the moving objects using movement. In other words, the system may move forward, backward, right or left to maintain capturing all targets in the scene. In this regard, the scene is divided into three major areas; unsafe area, safe area and over-safe area as shown in figure 1.

There are two parts inside the safe area; right side and left side, each of which has two parts. Totally, safe area has four different sub-areas; sLL (safe, left side of left side), sLR (safe, right side of left side), sRR (safe, right side of right side) and sRL (safe, left side of right side) as shown in figure 2.

There are also four different sub-areas inside unsafe area; uRR (unsafe, right side of right side), uLL (unsafe, left side of left side), uLR (unsafe, right side of left side), and uRL (unsafe, left side of right side) as shown in figure 2.

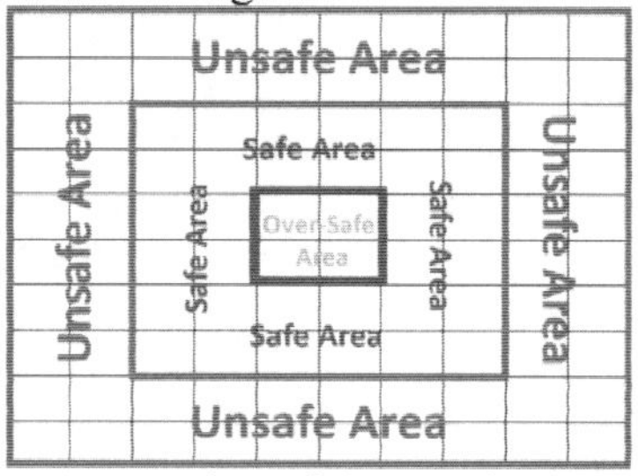

Figure 1. Location of three major partitions in scene

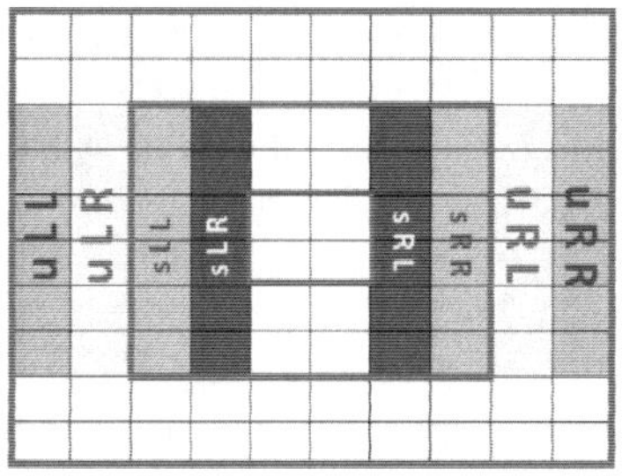

Figure 2. Different sub-areas inside unsafe area and safe area

Based on the partitioning, all movements of the system are based on two rules regarding their priorities;

R1) Keep all moving objects in the safe area, and

R2) Keep all moving objects in the center of the scene (not left side and not right side).

R1 that is always prior to R2, consists of two sub-rules regarding their priorities;

R1-a) Not to keep any moving object in unsafe area – for not losing the moving object, and

R1-b) Not to keep any moving object in over-safe area – for capturing and monitoring the target better, having more details.

R1 says when some moving object is inside unsafe area, there is a danger that the moving object gets out of the field of view, and in result the target gets lost, since the system cannot capture it anymore. On the other hand, when all moving objects are inside over-safe area, system moves towards the moving objects and gets closer to them such that the system captures and monitors the moving objects with more details.

When R1 is satisfied – no target is inside unsafe area and not all targets are inside over-safe area, R2 says if all moving objects are inside only one of the four sub-areas of safe area, then the system moves such that all targets are located inside the center of the scene. In fact R2 immigrates fairness to the system giving equal chance of monitoring to all moving objects that will come inside the scene from different directions.

Generally based on partitioning and location of the moving objects, the system can be at one and only one of three modes; unsafe mode, safe mode and over-safe mode.

4.3. Unsafe mode

Unsafe mode is the situation in which there exists at least one moving object in unsafe area. In this case, system has to move backward, move to right or move to left to follow the R1-a. In fact, the system using movement decreases the danger of losing the moving objects via not including them in unsafe area.

4.4. Safe mode

Safe mode is the situation in which there is no motion in the scene or there is at least one moving object in safe area while there is no moving object in unsafe area. System is said to be in SSNM (safe mode and no move) when;

1) The system is in safe mode and there exists at least one moving object in the left side of safe area and at least one target in the right side of safe area, or

2) The system is in safe mode and there is no moving object in the left side of safe area and there is no moving object in the right side of safe area.

4.5. Over-safe mode

The system is said to be in over-safe mode when the system is neither in unsafe mode nor in safe mode. In this situation, R1-a is already followed, and the system moves forward to follow R1-b. In fact, system increases the details of monitoring via moving forward and getting closer to the moving objects.

4.6. Unlimited backward movement

As mentioned earlier in video processing, all methods in motion detection including Gaussian filtering work based on comparing of the current video frame with one or more frames from the previous frame that is called as background. It means that, if there is some change in some pixels in two consequent frames, then the system considers this change as a motion. On the other hand, when system moves, almost entire scene changes and in result pixels of almost all three major areas (unsafe area, safe area and over-safe area) of consequent frames change. In this situation, RBMTM presumes to have motions in all areas including sub-areas of unsafe area. Hence, RBMTM proceeds to unsafe mode and has to perform backward movement to change the situation from unsafe mode to safe mode, but the result of movement is unsafe mode again. In other words, with first movement, RBMTM is fooled and gets inside an unlimited loop of unsafe mode and then backward movement.

To omit the problem of unlimited backward movement, RBMTM does not perform motions detection during its movement.

4.7. Algorithm

When RBMTM starts capturing and monitoring, value of the variable "target" is false, since the system did not detect any moving object yet. The function detect_motions is responsible to detect all moving objects among all areas in the scene. This function changes the value of the variable target to true if there is a motion. The function detect_motions also returns the area and sub-area of all moving objects. Then the system discovers its mode based on the areas and location of the moving objects. If the system is in unsafe mode, based on R1-a, system has to change the mode to safe mode using move_right, move_left, or move_backward.

If the system is in SSM, R1 is already followed, but based on R2, system has to change the mode to SSNM using move_right or move_left.

If value of variable target is true, and RBMTM is neither in unsafe mode nor in safe mode, then the system is in over-safe mode. In this situation, based on R1-b, system moves and gets closer to the moving objects using move_forward to change the mode to the safe mode.

The overall algorithm for system RBMTM is as following;

```
detect_motions()
while (target){
if (unsafe_state){if (uLL & NOT sRR & NOT
sRL){move_left twice;}
elseif (uRR & NOT sLL & NOT sLR) {
move_right twice;}
elseif (uLR & NOT sRR) {move_left;}
elseif (uRL AND NOT sLL) {move_right;}
else {move_backward;} exit;} // end of unsafe
if (safe_state){if ((sRR OR sRL) AND NOT
sLL AND NOT sLR){ move_right;}
elseif ((sLL OR sLR) AND NOT sRR AND
NOT sRL) {move_left;} exit;} // end of safe
move_forward; // over-safe sate
} // end of while
```

5. Implementation

From hardware view, the system RBMTM is a simple robot consisting a Raspberry Pi 3 Model B with a camera module v1, two servo motors (TowerPro SG5010) each of which connected to a wheel, power supplies (for Raspberry Pi and servo motors) and a free wheel in the front, while from software view, RBMTM uses Raspbian Jessie as operating system, OpenCV as library, and Python as programming language.

5.1. Partitioning

RBMTM uses ratios of width and ratios of height of the image (frame) to divide the scene to different areas and sub-areas. Each image of the scene is divided to many parts vertically based on height of the frame and horizontally based on width of the frame.

5.2. Motions detection

The system benefits Gaussian filtering to detect all moving objects. The system uses function "GaussianBlur", for Gaussian filtering, and then function "accumulateWeighted" and function "absdiff" to accumulate the weighted average between the current frame and previous frames due to calculate the difference between the current frame and running average frame. Then the system uses functions "threshold" and "dilate" in order to threshold the delta image and then dilate the thresholded image to fill the holes in the image based on the size of structuring element. Finally the system uses functions "findContours" and "contourArea" to find area and location of each moving object.

5.3. Movement unit

After detecting the areas and locations of moving objects, system may need to move to right, to left, forward or backward to satisfy R1 and R2. Since the system does not have any movement in Y-axis (vertical axis), the unit of movement is based on horizontal angle of view of the camera. As figure 1 and figure 2 show, the scene is basically parted into a 10×10-unit page, while horizontal angle of view of the camera is 53.50 [34]. Hence, the system uses the formula HMU=HAV/HBP, where HMU represents horizontal movement unit, HAV represents the horizontal angle of view, and HBP represents the horizontally basic partitioning of the scene. In result, the movement unit is equal to 5.35 degrees (53.50/10).

6. Result and discussion

Researchers tested the system in indoor. When there is no change in the light of indoor place (all lights are on) and the maximum number of moving objects is 6 – containing human, dog and electrical fan, while the maximum distance between moving object and the system is in the range of 0.5 to 5 meters, the best results are obtained when the resolution is at least set to 640×480, FPS is at least set to 16, minimum number of motion frames is set to 8, and minimum area of motion is set to 1500 pixels.

The system RBMTM will be tested in outdoor places where actual moving objects – such as moving cars and walking humans – exist to find the exact value of resolution, FPS, minimum area of motion and minimum number of motion frames.

References

[1] Pour Yousefian Barfeh D, Balba N, Bustamante R, Jose E, Albacea E and Mariano V 2017 Multi-informational auto tracking algorithm *I2BM,* Indonesia
[2] Tu Z and Bai X 2010 Auto-context and its application to high-level vision tasks and 3D brain image segmentation *IEEE Trans Pattern Anal Mach Intell*
[3] Belongie S, Malik J and Puzicha J 2002 Shape matching and object recognition using shape context *IEEE Trans. on PAMI* **24**(4) pp 509–522
[4] Oliva A and Torralba A 2007 The role of context in object recognition *Trends in Cognitive Sciences* **11**(12) pp 520–527
[5] Tu Z, Chen X, Yuille A and Zhu S 2005 Image parsing: unifying segmentation, detection, and object recognition *Int'l J. of Com. Vis.* **63**(2) pp 113–140
[6] Pitas L 2000 *Digital Image processing Algorithms and Applications*
[7] Wang L, Hu W and Tan T 2003 Recent Developments in Human Motion Analysis *Pattern Recognition* **36**(3)

IOP Conf. Series: Journal of Physics: Conf. Series **976** (2018) 012003 doi:10.1088/1742-6596/976/1/012003

[8] Lina H, Shafranb I, Yuhc D and Hagera G 2010 Towards automatic skill evaluation: Detection and segmentation of robot-assisted surgical motions *Taylor&francisOnline*. pp 220-230

[9] Caballeroa A, Castilloa J, Cantosc J and Tomasd R 2010 Optical flow or image subtraction in human detection from infrared camera on mobile robot *Robotics and Autonomous Systems* 58(12) pp 1273–1281

[10] http://www.axis.com/ph/en/products/ptz-cameras (Accessed August 22, 2017)

[11] https://www.bhphotovideo.com/c/product/752624-REG/Samsung_SPU_3750T_1_4_Vertical_Double_Density.html (Accessed August 22, 2017)

[12] http://www.bkgrupe.lt/uploads/docs/Hikvision/DS-2DF8836IV-AELW.pdf (Accessed August 22, 2017)

[13] Choi W and Savarese S 2012 A unified framework for multi-target tracking and collective activity recognition *12th European Conference on Computer Vision* Florence, Italy

[14] Butt A and Collins R 2013 Multi-target tracking by Lagrangian relaxation to min-cost network flow *CVPR* (2013)

[15] Andriyenko A, Schindler K and Roth S 2012 Discrete-continuous optimization for multi-target tracking *CVPR* (2012)

[16] Qin Z and Shelton C 2012 Improving multi-target tracking via social grouping *CVPR* (2012)

[17] Yang B, Huang C and Nevatia R 2011 Learning affinities and dependencies for multi-target tracking using a CRF mode *CVPR* (2011).

[18] Andriyenko A and Schindler K 2011 Multi-target tracking by continuous energy minimization". *CVPR* (2011)

[19] Benfold B and Reidstable I 2011 Multi-target tracking in real-time surveillance video *CVPR* (2011)

[20] Sridhar S, Mueller F, Oulasvirta A and Theobalt C 2015 Fast and robust hand tracking using detection-guided optimization *CVPR* (2015)

[21] Melax S, Keselman L and Orsten S 2013 Dynamics based 3d skeletal hand tracking *Canadian Information Processing Society*

[22] Xing J, Ai H and Lao S 2009 Multi-object tracking through occlusions by local tracklets filtering and global tracklets association with detection responses *CVPR* (2009).

[23] Jung B and Sukhatme G 2010 Real-time motion tracking from a mobile robot *International Journal of Social Robotics* **2**(1) pp 63–78

[24] Choi W, Pantofaru C and Savarese S 2012 Detecting and tracking people using an RGB-D camera via multiple detector fusion *International Workshops on Computer Vision* Daejeon, Korea

[25] Yang B and Nevatia R 2012 Multi-target tracking by online learning of non-linear motion patterns and robust appearance models *CVPR* (2012)

[26] http://www.axis.com/files/datasheet/ds_215ptz_34462_en_0902_lo.pdf (Accessed Aug 22, 2017).

[27] https://www.bhphotovideo.com/c/product/752624-REG/Samsung_SPU_3750T_1_4_Vertical_Double_Density.html (Accessed Aug 22, 2017).

[28] http://www.bkgrupe.lt/uploads/docs/Hikvision/DS-2DF8836IV-AELW.pdf (Accessed Aug 22, 2017).

[29] Shapiro L and Stockman G 2001 *Computer Vision*

[30] Nixon M and Aguado A 2008 *Feature Extraction and Image Processing*

[31] Ahi K 2016 Modeling of terahertz images based on x-ray images: a novel approach for verification of terahertz images and identification of objects with fine details beyond terahertz resolution *Proc. SPIE 9856, Terahertz Physics, Devices, and Systems X: Advanced Applications in Industry and Defense*

[32] Reinhard E 2006 *High dynamic range imaging: Acquisition, Display, and Image-Based Lighting* Morgan Kaufmann

[33] Fisher R, Perkins S, Walker A and Wolfart E 2003 *Spatial Filters - Laplacian of Gaussian*

[34] https://www.raspberrypi.org/documentation/hardware/camera/ (Accessed Aug 22, 2017).

Insertion Sorter in P Systems

Davood Pour Yousefian Barfeh[*,1,2,3] **Jonalyn G Ebron**[1] **and Jaderick P Pabico**[2]

[1]Lyceum of the Philippines University-Laguna, Calamba City, Philippines
[2]University of the Philippines-Los Baños, Laguna, 4031, Philippines
[3]Adamson University, 1000 Manila, Philippines

[*]davoodpb@gmail.com

Abstract. In this study researchers pay attention to the essence of Insertion Sort and propose a sorter in Membrane Computing. This research shows how a theoretical computing device same as Membrane Computing can perform the basic concepts same as sorting. In this regard, researches introduce conditional reproduction rule such that each membrane can reproduce another membrane having same structure with the original membrane. The researchers use the functionality of comparator P system as a basis in which two multisets are compared and then stored in two adjacent membranes. And finally, the researchers present the process of sorting as a collection of transactions implemented in four levels while each level has different steps.

1. Introduction

Membrane computing (MC) is a scope within computer science which explores to apperceive new computational models from the study of biological cells, particularly cellular membranes.

In 1998 when Gheorghe Păun initiated membrane computing [1], nobody was able to conceive this type of computing grows swiftly. The initial goal of membrane computing was to learn from biology especially cells something possibly useful to the field of computer science.

Păun officially proposed membrane computing in 2000 [2] and after that various types of membrane systems – known as P systems – were defined, all of them inspired from biological facts. Then numerous applications reported in several areas such as biomedicine, linguistics, computer graphics, economics, approximate optimization, and cryptography. Accordingly, several software products for simulating and implementing P systems were developed such as SNUPS (simulator of numerical P systems) and SRPSUMGPU (simulation of recognizer P systems by using Manycore GPU) [3–4].

2. Cell-like P systems

Cell-like P system – or for convenience we call them P system in this study – consists of many membranes arranged hierarchically. These membranes bound compartments. The compartments are the area where multisets of abstract objects are placed. The multisets are sets of objects (symbols) with multiplicities, while the objects are the chemicals in the compartments, swimming in some liquid such as water [5–6]. In other words, a multiset can be seen as a string in which multiplicity (number of each symbol) is significant, not the order of symbols. Since the objects are swimming and moving freely inside compartment, permutation or order of objects is not important [6–8]. For example consider multiset M1: abbbac, multiset M2: bacabb, and multiset M3: cbbaab. All of these multisets are the same, because in all of these multisets, number of object a is 2, number of object b is 3, and number of object c is 1. As it can be observed the orders of objects in multisets are not important since the objects are inside water and can freely move.

It is noticeable that, in each compartment there are some rules and the objects inside the compartment evolve according these rules. In other words, number of objects may change based on rules. The rules direct how the objects change or how the objects communicate across membranes [6–8]. Although, Gheorghe Păun and others proposed some desirable rules based on behaviour of cells, all of these rules pay attention to the objects inside membrane, while membrane itself is left behind. More specifically, there are situations in which a cell may reproduce another cell or a membrane may reproduce another membrane. This is why; researchers propose some rule regarding membrane reproduction. Totally, in this study, there are three types of rules. One of these types of rules makes changes in number of objects, while another type of rules changes the positions of the objects, and the last type of rules allows a membrane to reproduce. In the following, these three types of rules are explained in details.

2.1. Rules regarding change in number of objects

These rules change some object to some other object. For example, based on rule R1: ca → d, our previous multiset (as mentioned, all of 3 multisets M1, M2 and M3 are the same) changes to abbbd. As seen, one "a" and one "c" are transmuted to one "d". Now, based on rule R1, number of objects changed. Number of object "a" is 1, number of object "b" is 3 and number of object "d" is 1 (we do not talk about object "c" any more, since its number is zero).

2.2. Rules regarding change in position of objects

Some rules change the position of the object. In fact, this type of rules transfers the object from some membrane to some other membrane. Simple example about how the objects communicate across membranes, based on rules R2: a → ein and R3: d → foutg, our previous multiset (abbbd) changes to bbbg, since object "a" is converted to object "e" and object "e" is sent to a membrane located inside current membrane and object "d" is converted to object "f" and object "g", while the object "f" transfers to a membrane that covers the current membrane.

2.3. Rules regarding membrane reproduction

At some situations, membrane M can reproduce another membrane M such that the structure of new membrane M is exactly same as structure of the initial membrane M. Although we can have two types of reproduction - in and out, in this study we use the rule regarding reproduction-out.

When we use production-out rule for membrane M, another membrane M having same structure with the initial membrane M is generated out of the initial membrane M. Here the new membrane M may include the initial membrane M or not. In this study, new membrane M includes the initial membrane M. For example, there exists a membrane m0 containing two membrane m1 and m2 as shown in figure 1. If the rule reproduction-out is implemented on m0, then another membrane m0 containing m1 and m2 is generated out of initial membrane m0 while the new membrane m0 includes the initial membrane m0 as shown in figure 2.

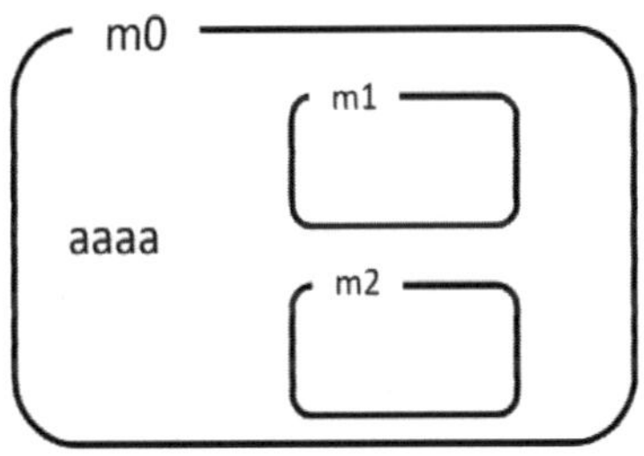

Figure 1. Initial membrane m0 before reproduction-out

As figure 2 shows, the structure of new membrane m0 (red color) is exactly same as the structure of the initial membrane m0 (black color). Here the structure means inclusion of other membranes, not inclusion of multisets. This is why; new membrane m0 does not include the multiset aaaa.

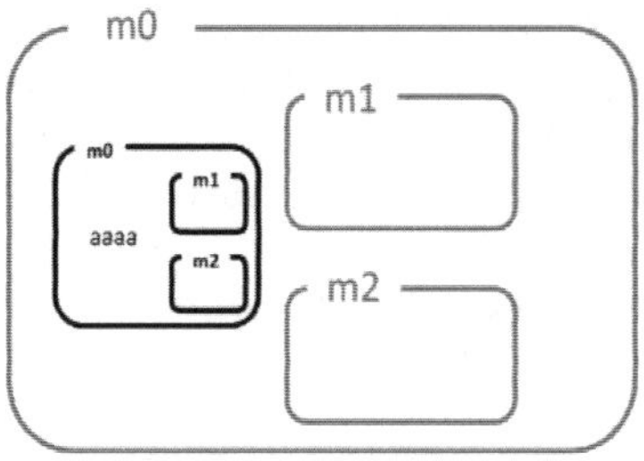

Figure 2. After reproduction-out; initial membrane m0 (black color) and new membrane m0 (red color)

The rules – belonging to all three types, can be implemented and applied in many ways. In general, rules emulate the biology in which chemical reactions performed in parallel as much as possible. In other words, transactions (rules) have maximal parallelism. There are another modes such as sequential, minimal parallelism and bounded parallelism. In sequential; one rule is used in each step, while in minimal parallelism; when a rule can be used, then at least one rule must be used, and in bounded parallelism; there is a restriction about the number of membrane or the number of rules to be used. In all modes mentioned, there is a common trait that is the objects and the rules regarding objects are chosen non-deterministically [1–6].

In membrane computing, a collection of transitions creates a computation. A computation generates a result as long as it halts that is to reach to a configuration where no rule can be applied [1–6].

3. Comparator P system

A comparator P system is able to sort two multisets as long as each multiset contains only one object, such as multisets s1: aaaaa and multiset s2: bbb. The comparator P system contains three membranes; m0, m1 and m2 while m0 covers m1 and m2 as shown in figure 3 [5].

At the beginning of the computation, two multisets which are a^x and b^y – where x is the number of object "a", and y is the number of object "b" – are in m0 while m1 and m2 are empty. Then all transactions are performed in two steps in order as following;

step 1:
  m0: ab → abin-m1
step 2:
  m0: a → bin-m2
  m0: b → bin-m2
  m1: b → bout

In step 1, the equal number of a`s and b`s transfer to m1, the rest of objects stay in m0. In step 2, three rules are performed in parallel. The rest of objects in m0 are converted to "b" and transfer to m2. Moreover all b`s in m1 move out from m1. These b`s transfer to m0, and then because of rule b → bin-m2 in m0, they transfer to m2. Now, halting state happens since no rule can be applied. In this case, the larger number (longer string) is in m2 and the smaller is in m1 [5].

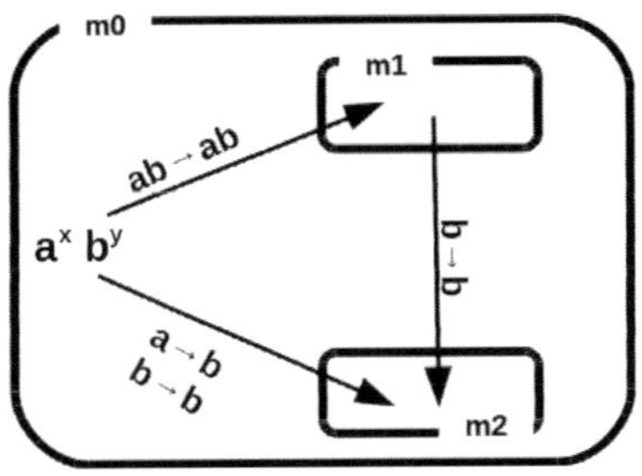

Figure 3. Comparator P system

4. Insertion sort

Insertion sort is a comparison-based algorithm in which the elements of the input list are sorted one at a time. In this algorithm, the sorted sub-list is always maintained in the lower position of the list. Then the new item is inserted into the previous sorted sub-list such that the new sub-list is also sorted.

We consider the list with one item as a sorted list. Then we iterate, considering one item of the list each repetition and growing the sorted sub-list. In other words, at each iteration, we work on an item from the input list and find its position in sorted sub-list comparing and swapping (if needed) the item with the elements of sorted sub-list started from the end. The iteration stops when swapping stops. The new sub-list is sorted and ready for the next iteration and new item from the list [9–12].

Insertion sort is slow compared to the advanced sorting algorithms such as quicksort, merge sort and heapsort [9]. Based on time complexity and number of comparison, insertion sort is as slow as bubble sort is; the worst case and the average case for both of insertion sort and bubble sort is O(n2) while the best case is O(n) where n is the number of elements of the input list. Although regarding time complexity, insertion sort is weak as bubble sort is, it has some strengths that make it bright despite bubble sort that is disregarded by some computer scientists [10–11]. The strengths of insertion sort can be listed as:

a) Simplicity: Jon Bentley used C programming language and implemented insertion sort only in three lines. He also implemented the optimized version of insertion sort in five lines [12].

b) Adaptiveness: insertion sort efficient for lists in which the elements are already considerably sorted. In this case the time complexity of insertion sort is O(nm) when each element in the input list is no more than m places far from its sorted position [9–12].

c) Stability: insertion sort does not change the relative order of items in the input list with equal indices [9–12].

d) In-placement: insertion sort needs only one extra unit memory space for swapping of two elements of the input list. The extra unit is small as size of elements of input list.

e) Online-ness: insertion sort sorts a list as it receives the elements of the list one-by-one [10–12].

Researchers appreciate these strengths as well as natural characteristic of insertion sort that is comparing two adjacent items then swapping them (if needed), and then implement insertion sort in comparator P system to propose a sorter in the field of membrane computing.

5. Insertion sorter in P systems

Now, researchers propose a membrane sorter containing comparator P system that uses insertion sort as algorithm to sort a list of multisets.

All membranes used in this sorter are in the form of comparator P system. When some comparator P system is reproduced, it is exactly a comparator P system which has 3 membranes; m0, m1 and m2 while m0 covers m1 and m2. Moreover all rules of new comparator P system are inherited from initial comparator P system.

The rules in insertion membrane-sorter has 4 levels which are performed in order inside each comparator P system, and each level has different steps which are also performed in order. At the beginning of the computation there is only one comparator P system while first and second multisets are inside m0 and following rules are applied;

level 1:

step 1:

m0: ab $\rightarrow$ $ab_{in\text{-}m1}$

step 2:

m0: a $\rightarrow$ $b_{in\text{-}m2}$

m0: b $\rightarrow$ $b_{in\text{-}m2}$

m1: b $\rightarrow$ b_{out}

level 2:

step 1:

m2: b $\rightarrow$ b_{out}

step 2:

m0: $reproduction_{out}$ | if there is no outer membrane

step 3:

m0: b $\rightarrow$ b_{out}

level 3:

m1: a $\rightarrow$ bout

m0: b $\rightarrow$ $a_{in\text{-}m0}$ | if there is inner m0

level 4:

outside: new multiset $\rightarrow$ $a\text{-}multiset_{in}$

At the start point; there exists only one comparator P system while two empty multisets m1 and m2 are inside m0.

Level 1 contains exactly the rules of a single comparator P system. At the end of this level, the greater number (multiset) is in m2 and the smaller is in m1 while m0 is empty.

In level 2, at step 1; the greater number transfers from m2 to m0. At step 2; if current m0 is the most outer membrane, then a new comparator P system is reproduced out of current m0 while new comparator P system includes current m0. At step 3; the greater number that contains only b's, transfers to outer membrane. At the end of level 2, the greater number is in m0 of outer comparator P system, while the smaller number is still in m1 of current comparator P system while m0 of current comparator P system is empty.

In level 3, the smaller number transfers from m1 to m0 while converting to b's. Moreover, if there is an inner m0, the number goes inside inner m0 to be compared with another number inside another comparator P system.

In level 4, if there is still some multiset in the list, it goes inside m0 of the outer comparator P system to be compared with the number that is inside m0 of the outer comparator P system.

It is notable that all levels can be performed in parallel but in different comparator P systems. For example, some comparator is working on level 2, and some other comparator is working on level 3, while another comparator might be idle.

The sorter system stops when there is no rule to be done. In this case, there is no multiset to go inside m0 of the outer comparator P system that is the rule in level 4, and all multisets are already sorted that are rules in levels 1, 2 and 3. When the system stops, the greatest number is in m0 of outer-most comparator and the smallest is in m0 of the inner-most comparator P system. Moreover, all m1's and all m2's are empty.

The proposed sorter uses natural characteristics of insertion sort in which compares two adjacent numbers (multisets) and then swaps them (if needed) in levels 1, 2 and 3. These comparisons and swaps are performed until the sub-list is sorted. Moreover, the sorter is working online same as insertion sort; when the new item arrives in level 4, the item finds its position with comparing and then swapping (if needed) started from the end of sub-list that is the outer-most comparator P system. In this way, at the end of each iteration, the sub-list is sorted that is the greatest number is in m0 of the outer-most comparator P system and the smallest number is in m0 of the inner-most comparator P system.

Acknowledgments
The researchers would like to extend profound gratitude and appreciation to the people from Lyceum of the Philippines University – Laguna, who contributed assistance to the fulfilment of this research paper, especially Dr. Peter Laurel – President, engineer Ricky Bustamante – Dean, College of Engineering and Computer Studies, and Mrs. Gerby Muya – Director, Research and Statistics Center.

References

[1] Păun G 1998 *Introduction to Membrane Computing*
[2] Păun G 2000 Computing with membranes *Journal of Computer and System Sciences* **61**(1) pp 108–143, and *TUCS* Report No **208**.
[3] Gutierrez–Naranjo M A, Perez–Jimenez M J and Riscos–Nunez A 2006 Available membrane computing software *Applications of Membrane Computing* Springer–Verlag, Chapter 15 pp 411–436
[4] Martinez–del–Amor M A, Perez–Hurtado I, Perez–Jimenez M J, Cecilia J M, Guerrero G D and Garcıa J M 2009 Simulation of Recognizer P Systems by Using Manycore GPUs *Membrane Computing: 10th International Workshop*
[5] Ardelean I, Ceterchi R and Tomescu A I 2008 Sorting with P Systems: A Biological Perspective. *Romanian Journal of Information Science and Technology* **11**(3) pp 243–252
[6] Tingfang W, Zhiqiang Z, Păun G and Pan L 2016 Cell-like spiking neural P systems. *ELSEVIER: Theoretical Computer Science* **623** pp 180-189
[7] Singh K, Deep K and Nagarb A K 2014 Cell-like P-systems based on rules of Particle Swarm Optimization. *ELSEVIER: Applied Mathematics and Computation* **246** pp 546-560.
[8] Chen K, Wang J, Li M, Ming J, and Peng H 2015 Cell-Like Fuzzy P System and Its Application of Coordination Control in Micro-grid *Springer: Bio-Inspired Computing - Theories and Applications*
[9] Bender M, Farach - Colton M and Mosteiro M 2006 Insertion sort is O(n log n) *Journal of Theory of Computing Systems* **39**(3) pp 391–397
[10] Knuth D 1998 *The Art of Computer Programming* Volume 3: Sorting and Searching, Second Edition. Addison-Wesley
[11] Owen A 2003 Bubble sort: an archaeological algorithmic analysis *ACM SIGCSE Bulletin*
[12] Bentley J 2000 Programming Pearls *ACM Press* Addison–Wesley

IOP Conf. Series: Journal of Physics: Conf. Series **976** (2018) 012005 doi:10.1088/1742-6596/976/1/012005

Adjusting Beliefs via Transformed Fuzzy Priors

T Rattanadamrongaksorn[1], D Sirikanchanarak[1,3], J Sirisrisakulchai[2] and S Sriboonchitta[2]

[1] Faculty of Economics, Chiang Mai University, Chiang Mai 50200, Thailand
[2] Puey Ungphakorn Center of Excellence in Econometrics, Faculty of Economics, Chiang Mai University, Chiang Mai 50200, Thailand
[3] Bank of Thailand, Bangkok 10200, Thailand

tanarat_ra@cmu.ac.th, http://orcid.org/0000-0002-4401-0011

Abstract. Instead of leaving a decision to a pure data-driven system, intervention and collaboration by human would be preferred to fill the gap that machine cannot perform well. In financial applications, for instance, the inference and prediction during structural changes by critical factors; such as market conditions, administrative styles, political policies, etc.; have significant influences to investment strategies. With the conditions differing from the past, we believe that the decision should not be made by only the historical data but also with human estimation. In this study, the updating process by data fusion between expert opinions and statistical observations is thus proposed. The expert's linguistic terms can be translated into mathematical expressions by the predefined fuzzy numbers and utilized as the initial knowledge for Bayesian statistical framework via the possibility-to-probability transformation. The artificial samples on five scenarios were tested in the univariate problem to demonstrate the methodology. The results showed the shifts and variations appeared on the parameters of the distributions and, as a consequence, adjust the degrees of belief accordingly.

1. Introduction

There exist the situations that changes in important factors such as technological constraints, individual preferences, market conditions, etc. occur profoundly and permanently which are called structural changes and cause the shifts of the properties of data. Thus, the inference and prediction by the existing parameters are unreliable. Fortunately, the estimation based solely on human intuition could be made by the so-called expert opinion. For example in, but not limited to, the financial application, as soon as the new tax law is enacted, the exporter raise their price even before they really pay for the new cost. As a matter of fact, the calculation in backgrounds such as labor wages, material prices and other related costs are complex and time-consuming but the expert can neglect those details and pursue with their approximation coming out of the thin air. Most of the time, surprisingly, their expert opinions are reasonably correct. We therefore propose the incorporation of the existing historical observations with this subjective data which are handled well by fuzzy set theory [1].

The fuzzy set theory is the generalization of set theory to represent the non-random uncertainty. Fuzzy set is characterized by a set of pairs—the elements and their possibilities, $[\theta, \pi(\theta)]$. The element, θ, is a considered parameter while its pair is the possibility of occurring in the specific value. The concept is efficient in dealing with the linguistic expressions by fuzzy number which is the representation of quantity that cannot be described precisely due to the incomplete knowledge.

2. Representing linguistic terms by fuzzy priors

Rather than specifying the knowledge about data by probability distributions, it is much easier and more practical for the person in the field to convey the messages about something verbally such as "at least", "most positive", "exactly", etc. In this study, we therefore translate the linguistic terms into mathematical expressions by fuzzy numbers. Fuzzy number is the mathematical expression of the quantity or parameter that takes the form of $N = (a, b, c, d)$ whose possibilities $(0,1,1,0)$ are usually omitted from writing. The translation of the linguistic terms into fuzzy representations can be achieved by defining the fuzzy numbers that has been adapted from and introduced in the previous works [2,3]. Currently, our collection of fuzzy representations grows from the original work to the list in Figure 1. Because later we integrate these representations into Bayesian framework as priors, we may call them as fuzzy prior.

The examples of fuzzy priors for our experiment are 1) No distribution, 2) Uniform, 3) Mean-Variance, 4) Interval, and 5) Left-skewed and the parameters for generating statistical data in normal model are shown in Table 1.

Table 1. Fuzzy and statistical properties of the artificial samples.

Name	Largest possible	Most possible	$\theta-$centroid[a]	$\pi-$centroid[a]	Mean	SD	Size
No distribution	50	50	50	1.00	0	5	100
Uniform	[20,80]	[20,80]	50	1.00	0	5	100
Mean-Variance	[20,80]	50	50	0.50	0	5	100
Interval	[20,80]	[40,60]	50	0.33	0	5	100
Left-skewed	[20,80]	60	53	0.50	0	5	100

[a] $\theta-$centroid, $\tilde{\theta} = \int \theta\, \pi(\theta)d\theta / \int \pi(\theta)d\pi$, and $\pi-$centroid, $\tilde{\pi} = \int \theta\, \pi(\theta)d\pi / \int \pi(\theta)d\pi$, are the expectations of fuzzy numbers on the corresponding axes but the latter is not applicable in this work.

There are, however, the incompatibilities between the usages of expert opinions and historical observations because they are based on the different theories and, on the other hand, utilizing only either of them is loss of opportunity. Taking a closer look into the real-life event, we often find that even the data are structural changed but their behavior may remain almost unchanged. For instance, the expected price of a stock has been elevated due to the new appointment of its administrative team but the variation is not altered because of unchanged market conditions. This implies that while there are needs for the new evaluation of data, the existing knowledge should not be neglected. The better alternative is to combine both data via, we propose, the possibility-to-probability transformation.

IOP Conf. Series: Journal of Physics: Conf. Series **976** (2018) 012005 doi:10.1088/1742-6596/976/1/012005

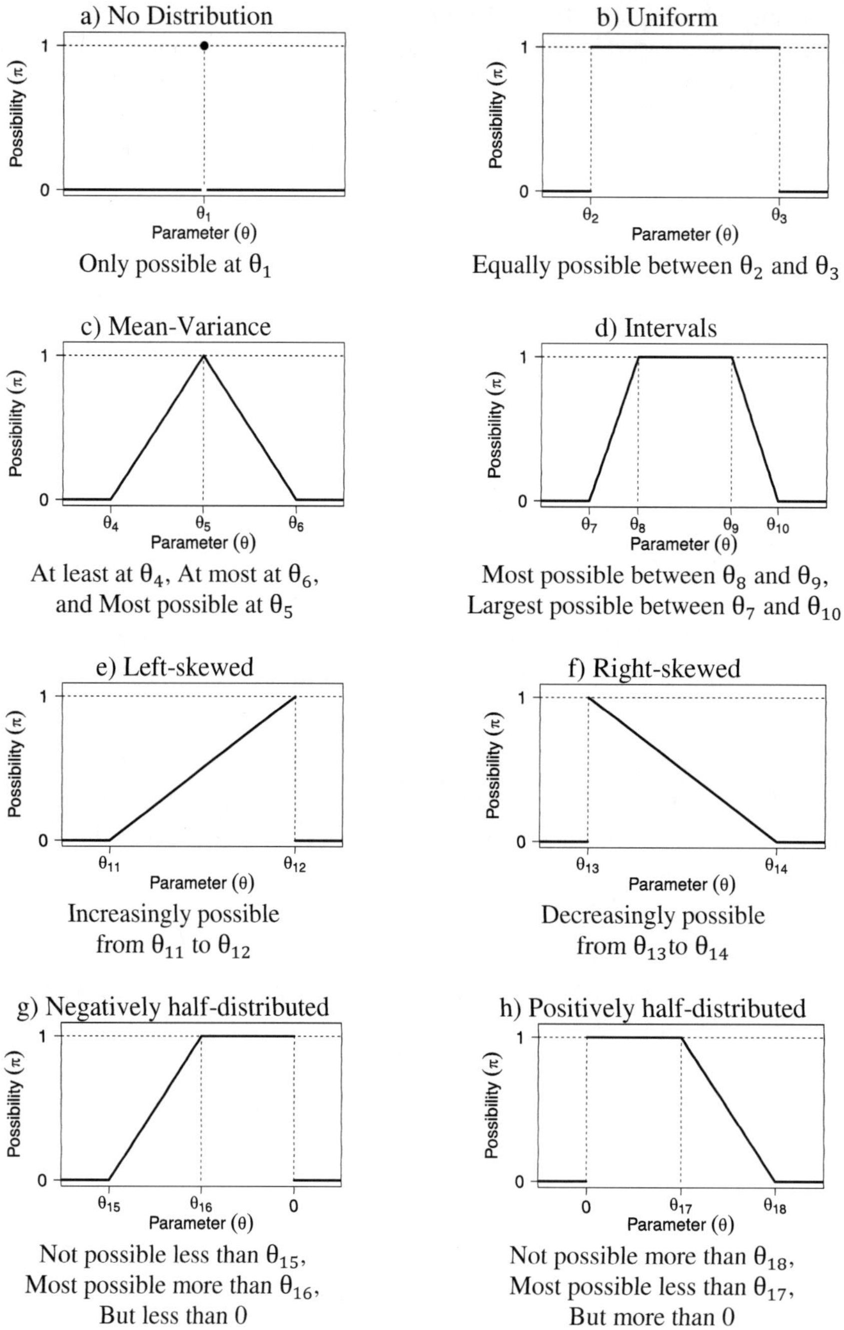

Figure 1. Fuzzy representations

3. Transforming expert opinions to probability distributions

In this study, we consider the transformation exclusively from the possibility to probability due to the loss of information from the inverse operation. In order to obtain the probability distribution from fuzzy number, the transformation is activated for M cycles by: 1) randomly select the possibility from the uniform distribution which is called alpha-cut fuzzy number at the selected possibility to obtain a new fuzzy number, A_α, 2) randomly choose the possibility, π_a, from the first step, 3) sort the results in descending order, 4) repeat the above steps for M times, and 5) calculate the transformed probabilities by [4]:

$$P_i(x) = \sum_{j=i}^{M-1} \frac{\pi_j(x) - \pi_{j+1}(x)}{j}, \tag{1}$$

where $max(\pi_a) > \pi_1 > \pi_2 > \ ...\pi_j ... > \pi_{M-1} > \pi_M = min\,(\pi_a)$. The possibilities are transformed by accumulating the differences between the consecutive possibilities and divided by its decreasing order of transformation, j, from the current (order i) to the one before the last order, $M-1$. As a result, the distribution shapes of the transformed probabilities will gradually increase from the start on both sides with acceleration and reach its peak at the highest probability.

The results of transformations are depicted in the left column of Figure 2. The transformed probabilities in the case of no distribution and uniform are unchanged due to the fact that the estimations are definitely possible which are normally the rare cases, however. The remaining cases show the changes on the side of the distributions which are lower in transformed probabilities than those in possibilities but the shapes still follow their fuzzy priors. On the sides of fuzzy numbers, the concave appeared and noticeably more curved in the triangular fuzzy numbers (Mean-Variance and Left-skewed) than those in the trapezoid (Interval) because more possibilities can be identified thus induced more probabilities.

4. Updating beliefs by Bayes' theorem

Bayesian approach is a statistical method that assumes randomness in parameter. It is attractive in its updating mechanism through the Bayes' theorem:

$$P(\theta|X) = \frac{P(\theta)P(X|\theta)}{P(X)}, \tag{2}$$

where θ and X represent parameters and data respectively. The equation indicates that the conditional probability of parameter given data or *Posterior*, $P(\theta|X)$ is proportional to the product of the initial belief in the probability of interested parameters or *Prior*, $P(\theta)$ and the joint probability of data or *Likelihood*, $P(X|\theta)$ and normalized by the Marginal Likelihood, $P(X)$—which is usually ignored in computation. The formula can be interpreted that as the evidence is available, the prior is updated by or combined with the probability of data that yields in the probability of posterior.

The results of our experiments shown in Figure 2 and Table 2 were obtained by the Markov-Chain Monte-Carlo (MCMC) simulation. The posteriors were developed from the mixtures of likelihoods and priors. The means of the distributions (50.04-59.75) stand approximately on the centroids (50-53) of the fuzzy priors. The shapes of distributions were developed from the original shapes of the priors but deteriorated on the sides close to the data. The ranges of distributions are close to the ranges of fuzzy priors that the minimums (20.21-25.25) and maximums (76.10-79.89) are around the lower and upper bounds of fuzzy priors (20 and 80 respectively). The exception is on the case of no-distribution that has a little of variation causing merely by the variance of randomness, $u \sim Unif(0,1)$, in the sampling algorithm. Both illustrations and numerical results confirm the significant impacts from the expert opinions by the differences from those by the MLE obviously seen in the figure and table, even though the numerical results are not included.

IOP Conf. Series: Journal of Physics: Conf. Series **976** (2018) 012005 doi:10.1088/1742-6596/976/1/012005

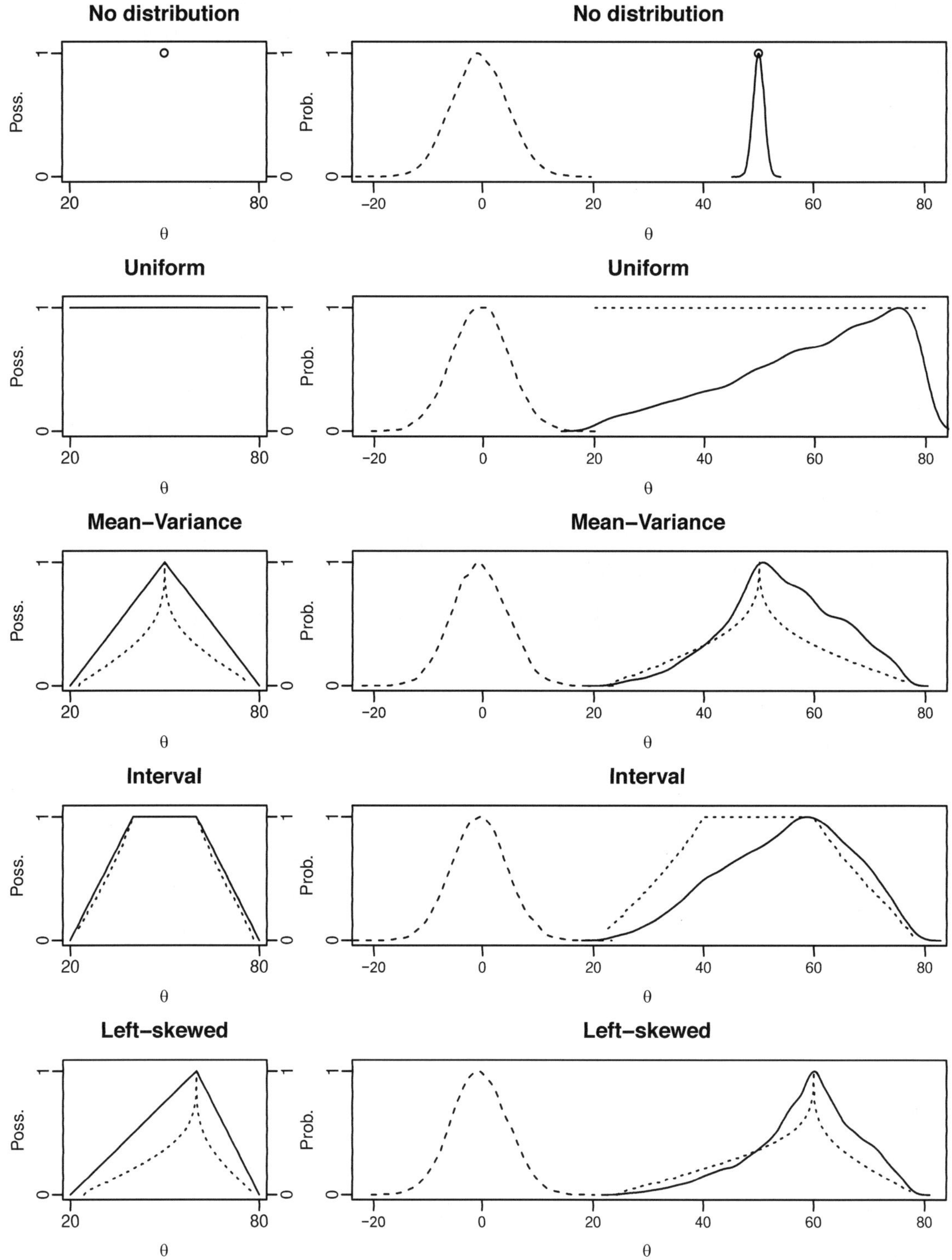

Figure 2. Left – Priors' Fuzzy (circle or solid) numbers and Transformed (left, circle or dotted) probability distribution. Right – Results from Maximum Likelihood Estimation (dashed) vs Transformed (circle or dotted) and Posterior (solid) probability distributions by the proposed method.

Table 2. Statistics of the estimates by the MLE and the posteriors by the proposed method.

Method	Mean	SD	Min.	Max.	Median	Mode	C.I. / C.R.[b]	Size
MLE	-0.57	4.96	-14.46	13.34	0.08	0.59	[-10.03,9.10]	100
Proposed method								
No distribution	50.04	1.00	45.64	53.62	50.04	50.03	[48.08,51.96]	10000
Uniform	59.75	14.64	20.21	79.89	62.58	75.02	[26.82,79.07]	10000
Mean-Variance	53.98	10.20	23.30	76.10	53.58	50.63	[33.05,73.11]	10000
Interval	54.94	11.40	22.74	78.03	56.08	58.76	[31.66,74.11]	10000
Right-skewed	58.25	9.45	25.25	77.30	59.41	60.06	[36.30,74.14]	10000

[b] C.I. and C.R. are abbreviations for Confidence Interval (for MLE) and Credible Interval (for the proposed method) respectively.

5. Conclusions

In the situation that the statistical analysis is insufficient, we propose the fusion of the expert opinions and historical data through the Bayes' updating mechanism with the possibility-to-probability transformation. The proposed methodology includes 1) Representing linguistic terms by fuzzy priors, 2) Representing linguistic terms by fuzzy priors, and 3) Updating beliefs by Bayes' theorem.

With the artificial samples, we were able to initiate the various cases of the initial knowledges. The benefit was that the results from the different formulations could easily be traced and understood. The posterior probabilities from the experiments appeared to be compromised between the prior knowledge and the past observations. The outcomes from our experiments were biased by the priors but varied by the likelihoods. In simpler terms, the positions of the results were indicated by the expert opinions but their shapes were influenced by the distributions of the existing data.

This research contributes to the acquisition and utilization of subjective data within Bayesian framework. The method is more natural and easier than the traditional implementations by the expert opinions that can be translated from linguistic terms into fuzzy numbers which are intuitive and flexible to users. However, it also pays the price in computation time and its accuracy depends very much on the quality of expert opinions. The method is so general that can be adapted to the applications in various disciplines that require the intervention of and adjustment by human. The concurrent [5] and future research include the more practical real-life situations like multivariate inference and optimization which involve much complex and require more sophisticate techniques.

Acknowledgments
The first author thanks to Dr. Supanika Leurcharusmee and Asst. Prof. Dr. Napat Harnpornchai for their kind discussions on some ambiguities arising during the problem formulation.

References

[1] Zadeh L 1965 Fuzzy sets *Inf. Control* **8(3)** pp 338–53
[2] Lorterapong P and Moselhi O 1996 Project-network analysis using fuzzy sets theory *J. Constr. Eng. Manag.* **122(4)** pp 308–18
[3] Rattanadamrongaksorn T, Sirisrisakulchai J and Sriboonjitta S 2016 Usages of Fuzzy Returns on Markowitz's Portfolio Selection *Lecture Notes in Computer Science Integrated Uncertainty in Knowledge Modelling and Decision Making* pp 124–35
[4] Dubois D, Prade H and Sandri S 1993 On possibility/probability transformations Fuzzy Logic: State of the Art *Springer Netherlands* Dordrecht pp 103–12
[5] Rattanadamrongaksorn T, Sirikanchanarak D, Sirisrisakulchai J and Sriboonchitta S 2017 Adjusting beliefs via transformed fuzzy prices — the manuscript submitted to The First International Econometric Conference of Vietnam: ECONVN2018

Chinese Text Summarization Algorithm Based on Word2vec

Xu Chengzhang and Liu Dan

Research Institute of Electronic Science and Technology, University of Electronic Science and Technology of China, Chengdu 611731, China

15208171819@163.com

Abstract. In order to extract some sentences that can cover the topic of a Chinese article, a Chinese text summarization algorithm based on Word2vec is used in this paper. Words in an article are represented as vectors trained by Word2vec, the weight of each word, the sentence vector and the weight of each sentence are calculated by combining word-sentence relationship with graph-based ranking model. Finally the summary is generated on the basis of the final sentence vector and the final weight of the sentence. The experimental results on real datasets show that the proposed algorithm has a better summarization quality compared with TF-IDF and TextRank.

1. Introduction

With the advent of the era of big data, the volume of data on the Internet is growing rapidly. This makes artificial text summarization methods unable to meet the needs of users. Text summarization technology is one of the effective tools to solve this problem. Text summarization technology is a method that can extract or generate the summarizations under the help of computers [1].

There are two main methods to get the summary, the extracted summarization method and the generative summarization method.

The extracted summarization method assumes that the topic can be summed up into some sentences or a few words in the text, so the text summarization problem can be transformed into a sentence representation and sorting problem. In the literature [2], author proposed a statistical summarization method. TF-IDF algorithm was used to calculate the weight of words and sentences, and the summary was extracted from the text after sorting the sentences by weight. In the literature [3], author proposed a summarization method based on similarity. From the multi-document model, a probability distribution of words that indicates the relationship between words and topics was generated. The weight of sentence was calculated based on the similarity between words. Thus, the summary was extracted from the text after sorting the sentences. Moreover, extracted summarization methods based on machine learning [4] and graph model [5] were also effective in representing and sorting the sentences. In most of the above methods, the frequency statistics for each word were used to calculate the weight of the sentence and the sentences were sorted on the basis of the weight, the graph model and the text feature.

The generative summarization method is a method to generate the natural language that can cover the main topic of the article after the computer understands the content. In the literature [6], author proposed a data-driven approach. This approach utilized an attention-based model to generate each word of the summary. The model could easily be trained end-to-end and scaled to a large amount of training data. However, this method was limited by natural language processing technology and did not work well in long articles.

The text summarization algorithm used in this paper belongs to the extracted summarization method.

2. Text summarization algorithm based on Word2vec

FIG. 1 is the flowchart of the summarization algorithm based on Word2vec. The algorithm firstly divides the text into sentences and divides each sentence into words, and each word is represented as a vector trained by Word2vec. Then the algorithm calculates the weight of each word, the weight of each sentence and the sentence vector iteratively on the basis of the word vector, word-sentence relationship and graph-based ranking model. Finally, a summary is generated on the basis of the final sentence vector and final weight of the sentence.

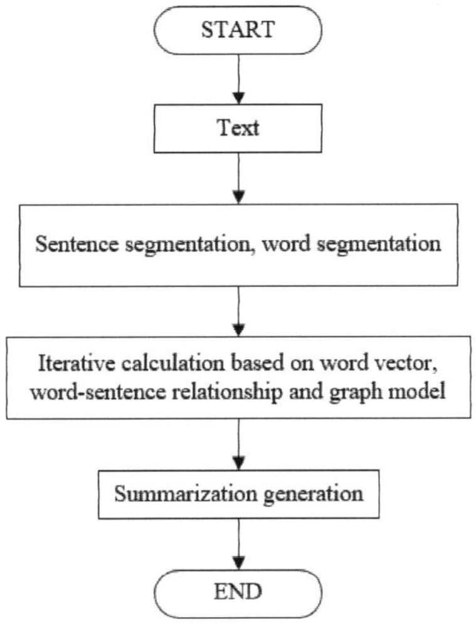

Figure 1. Flowchart of the summarization algorithm

2.1. Word2vec

Word2vec [7] is an open source toolkit to produce word vectors. Vectors trained by Word2vec have the following characteristics: the correlation between words can be measured by the vector distance between word vectors. The higher the semantic relevance is, the shorter the distance between two word vectors will be. Word2vec is now widely used in natural language processing tasks such as clustering, synonyms, Sentiment Analysis and text classification.

Word2vec can utilize Continuous Bag-of-Words Model (CBOW) and Skip-gram Model to produce word vectors. The main idea of CBOW is to predict the probability of the centre word on the basis of the context around the word, and Skip-gram Model is to "skip some symbols" around the centre word to predict the context [8]. While CBOW Model is time-consuming, and the training effect is limited by the size of the sliding window, Skip-gram Model has higher semantic accuracy, but the computational complexity is high too and takes a long time for the training process.

In Chinese text summarization, there are some algorithms which can consider the meaning of words and similarity between words [9] [10] [11], and most of these methods sort the sentences by text feature so that they are confined to certain texts and have no universal applicability. On the basis of Word2vec, the text summarization algorithm can to some extent preserve the semantic relevance between sentences.

2.2. Iterative calculation process based on word-sentence relationship and graph model

The basic idea of iterative calculation process based on word-sentence relationship and graph model is that on the one hand the more key words a sentence consists of, the higher weight this sentence will have. On the other hand, the more frequently a word appears in a high weight sentence, the higher weight this word will have.

The algorithm first defines the variables. Assuming that the set of sentences is *S*, the number of sentences is m, and the weights of the sentences are expressed as a m-dimensional vector $T = [T_1, T_2, ..., T_m]$, in which T_i represents the weights of the i-th sentence; the set of words is *W*, the number of words is *n*, and the weights of words are expressed as a n-dimensional vector $Y = [Y_1, Y_2, ..., Y_n]$, Y_i is the weight of the i-th word, furthermore, the initial weight of each word is calculated by TF-IDF [12] or assigned equally; Word2vec vector corresponding to *W* is $V = [V_1, V_2, ..., V_n]$; p_{ij} represents the number of occurrences of the word j in sentence i.

D represents words in a sentence, and the sentence is represented as a sentence vector x:

$$x = \sum_{i \in D} V_i \times Y_i \tag{1}$$

The similarity between sentences is calculated by using the cosine value of the angle between them.

$$sim(X,Y) = \frac{\vec{x} \cdot \vec{y}}{\|\vec{x}\| \cdot \|\vec{y}\|} \tag{2}$$

An undirected graph G is established in which the sentence is treated as node and the similarity between sentences is treated as edge. After several iterative computations combined with word-sentence relationship and PageRank [13] on G, the final weight of each word, the final weight of each sentence and the final sentence vector can be calculated. Experiments in the literature [5] proved that after several iterations of the weights, the weight of sentence converged. The terminating condition is a threshold. When the threshold is 0.0001, after 20-30 iterations, the weight value converges [14]. To get the summary, sentences which have a higher similarity to the sentence with the highest weight than a threshold are regarded as repetitions and omitted. Ultimately, summaries can be generated in the top K sentences, and the K can be determined according to some conditions like the number of summaries and the number of words in summary /the number of text words.

2.2.1. Iterative computation . The iterative computation is a combination of word-sentence weight-propagation and PageRank. Based on the word frequency in a sentence, there is a weight-propagation between words and sentences. Word2vec is used to further update the vector representation of the sentence.

The iterative computation firstly runs a sentence weight calculation algorithm like PageRank on G:

$$T_i = \frac{1-d}{m} + d \times \sum_{d \in In(d_i)} T_j \times \frac{sim(d_i, d_j)}{\sum_{d_k \in Out(d_j)} sim(d_j, d_k)} \tag{3}$$

In where, d is the damping coefficient (default is 0.85), m can be 1 or the number of the sentences in the article. The impact of other sentences on the current sentence is controlled by d and m. d_i is a sentence node in graph G, which represents the i-th sentence in S. $In(d_i)$ represents the set of nodes pointing to the sentence node d_i and $Out(d_j)$ represents the set of nodes pointed to by d_j . T_j is the weight of the sentence node d_j in this round. In (3), the weight of the current sentence is calculated on the basis of the similarities to other sentences and the weight of other sentences. Perform such a calculation for each sentence until the weight converges and then we can get the initial weight of the sentence.

Based on the weights of sentences in (3), the weights of words can be updated:

$$Y_j = \left(\sum_{i=1}^{m} p_{ij} \times T_i\right) / \left(\sum_{i=1}^{m} p_{ij}\right) \tag{4}$$

After the weight of each word is obtained, the weight of each sentence can be recalculated and the sentence vector can be recalculated by (1):

$$T_i^{'} = (\sum_{j=1}^{n} p_{ij} \times Y_j) / (\sum_{j=1}^{n} p_{ij}) \tag{5}$$

Finally, after a combination of formula (3), (4) and (5), the weight of words and sentences, and sentence vector can be the initial value in the next iteration, α is the adjustment factor, the sentence update method is:

$$T_i = a \times T_i + (1 - a) \times T_i^{'} \tag{6}$$

After several iterations, algorithm will get the final weight of sentence and the final sentence vector.

2.2.2. Summarization filter. Based on the final weight of each sentence and the final sentence vector, sentences which have a higher similarity to the sentence with the highest weight than a threshold are regarded as repetitions and omitted. To generate the summary, number of words in summary / number of text words is considered as the precondition for K, and the summary is the remaining top-k sentences.

3. Experiment

3.1. Experimental corpus

There is lack of authoritative test data in the field of Chinese text summarization. By using web crawler, some news from Ifeng that belongs to different subject like finance, entertainment, culture and sports were fetched. On the fetched data a Word2vec model was trained too. The experiment was conducted based on 100 news data from all areas mentioned above. The statistical characteristics of our experimental data are shown in Table 1. In our experimental data, there were 2722 sentences and 149,668 words in total. The minimum number of words in an article was 365, the maximum number of words was 6425; the minimum number of sentences was 4, and the maximum number of sentences was 149. To generate the summary, number of words in summary / number of text words was considered as the precondition for K. The weight of the sentence was sorted in advance artificially, and the top-K sentences were selected as the expert summaries. Edmundson[15] was used to evaluate the results. Edmundson was defined as:

$$\text{Coincidence degree} = \frac{\text{Number of matched sentences}}{\text{Number of expert summaries}} \times 100\% \tag{7}$$

Table 1. Statistical characteristics of news corpus

Number of news	100
Total number of sentences	2722
The maximum number of sentences in an article	149
The minimum number of sentences in an article	4
Total number of words	149668
The maximum number of words in an article	6425
The minimum number of words in an article	365

3.2. Analysis of experimental result

In the experiment, the precondition for K was that number of words in summary / number of text words was close but no more than 10%. The experimental result is shown in Table 2. The coincidence degree in financial news was 34.06%, in entertainment news was 31.23%, in sports news was 38.29%, and in culture news was 27.44%. Respectively, the average was 31.54%. The result shows that the effect of Chinese text summarization algorithm based on Word2vec is better in financial, entertainment, sports news, and is average in cultural news. In the financial, entertainment and sports test data, the number of words contained in a sentence is smaller, so the Word2vec-based feature representation method can better retain the relevance between the sentences. In addition, compared with entertainment and culture news, there are more repetitive keywords and stronger semantic relevance in financial and sports news, which make the weights of sentences more accurate.

Table 2. Experimental results

Category	**Average of K**	**Coincidence degree /%**
Finance	4.3	34.06
Entertainment	1.82	31.23
Sports	2.46	38.29
Culture	6.75	27.44
Average	3.14	31.54

3.3. Comparison with the original method

On the same experimental corpus, the proposed method was compared with TF-IDF and TextRank, and the evaluation result is shown in Figure 2. Compared with TF-IDF and TextRank, the method proposed in this paper had the highest coincidence degree in financial, entertainment and sports news, while in cultural news the coincidence degree was only almost 1.6% lower than that of TextRank. On average, the coincidence degree of our method was 6-7% higher than TF-IDF and TextRank. The result shows that method proposed in this paper has a better summarization quality in news from most subjects compared with TF-IDF and TextRank.

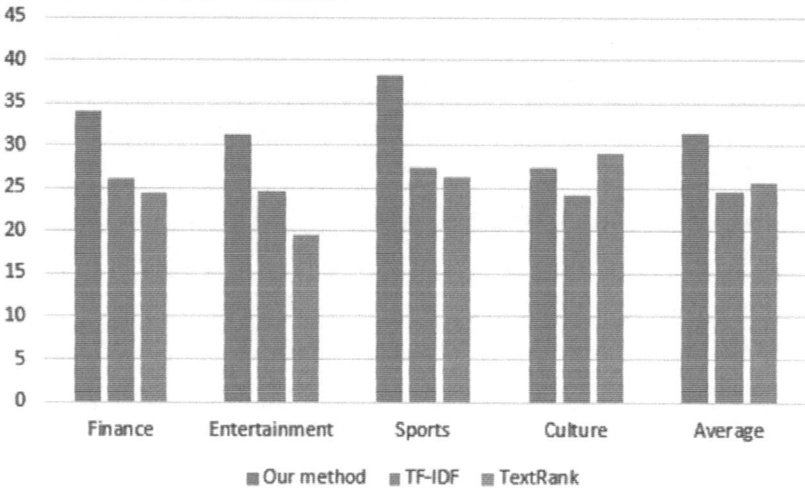

Figure 2. Result of comparative Experiment

4. Conclusion

Focusing on the issue that Chinese text summarization needs to extract some sentences to cover the topic of an article, we proposed a Chinese text summarization algorithm based on Word2vec. After converting words into vectors, the algorithm updates the weight of words and sentences iteratively on the basis of the combination of word-sentence relationship and the graph ranking model. Experimental results on real datasets shows that the text summary algorithm has a better summarization quality compared with TF-IDF and TextRank.

References

[1] Tan Chong, Chen YueXin. Literature Review of Automatic Summarization Method[J]. Journal of The China Society for Scientific and Technical Information, 2008, 27(1):12-12.
[2] Cheng QianQian, Tian DaGang. Automatic Chinese Summarization Model Based on Basic Elements Method[J]. New Technology of Library and Information Service, 2010, 26(2):74-78.
[3] Wang Hongling, Zhang MingHui, Zhou GuoDong. Chinese multi-document summarization system based on topic information[J]. Computer Engineering and Application, 2012, 48(25):132-136.
[4] Cao Yang, Cheng Ying, Pei Lei.A Review on Machine Learning Oriented Automatic Summarization[J]. LIBRARY AND INFORMATION SERVICE, 2014, 58(18):122-130.
[5] Mihalcea R, Tarau P. TextRank: Bringing Order into Texts[J]. Unt Scholarly Works, 2004:404-411.
[6] Rush A M, Chopra S, Weston J. A Neural Attention Model for Abstractive Sentence Summarization[J]. Computer Science, 2015.
[7] Mikolov T, Sutskever I, Chen K, et al. Distributed Representations of Words and Phrases and their Compositionality[J]. Advances in Neural Information Processing Systems, 2013, 26:3111-3119.
[8] Xiong FuLin, Deng YiHao, Tang XiaoCheng. The Architecture of Word2vec and Its Applications[J]. Journal of Nanjing Normal University, 2015(1):43-48.
[9] Huang ChengHui, Ying Jian, Hou Fang. A Text Similarity Measurement Combining Word Semantic Information with TF-IDF Method[J]. Chinese Journal of Computers, 2011, 34(5):856-864.
[10] Cheng Yuan, Wushouer SILAMU, Maimaitiyiming HASIMUA. Automatic Text Summarization Based on Comprehensive Charateristics of Sentence[J]. Computer Science, 2015, 42(4):226-229.
[11] Jiang XiaoYu. Automatic Summarization Algorithm Based On Keyword Extraction[J]. Computer Engineering, 2012, 38(03):183-186.
[12] Tang Ming,Zhu Lei, Zou XianChun. Document Vector Representation Based on Word2Vec[J]. Computer Science, 2016, 43(6):214-217.
[13] Page L. The PageRank citation ranking : Bringing order to the web[J]. Stanford Digital Libraries Working Paper, 1998, 9(1):1-14.
[14] Luhn H P. The automatic creation of literature abstracts[M]. IBM Corp. 1958.
[15] Edmundson H P. New Methods in Automatic Extracting.[J]. Journal of the Acm, 1969, 16(2):264-285.

IOP Conf. Series: Journal of Physics: Conf. Series **976** (2018) 012007 doi:10.1088/1742-6596/976/1/012007

Data Prediction for Public Events in Professional Domains Based on Improved RNN- LSTM

Bonan Song, Chunxiao Fan, Yuexin Wu and Juanjuan Sun

School of Electronic Engineering, Beijing University of posts and telecommunications, Beijing, China

songguolan1@163.com, cxfan@bupt.edu.cn, wuyuexin@bupt.edu.cn, sunjj@bupt.edu.cn

Abstract. The traditional data services of prediction for emergency or non-periodic events usually cannot generate satisfying result or fulfill the correct prediction purpose. However, these events are influenced by external causes, which mean certain a priori information of these events generally can be collected through the Internet. This paper studied the above problems and proposed an improved model—LSTM (Long Short-term Memory) dynamic prediction and a priori information sequence generation model by combining RNN-LSTM and public events a priori information. In prediction tasks, the model is qualified for determining trends, and its accuracy also is validated. This model generates a better performance and prediction results than the previous one. Using a priori information can increase the accuracy of prediction; LSTM can better adapt to the changes of time sequence; LSTM can be widely applied to the same type of prediction tasks, and other prediction tasks related to time sequence.

1. Introduction

The growing amount of data greatly enriched the knowledge base of data; a lot of potential research values are embedded in these data. Domains such as electricity, transportation and telecommunications as deeply influenced by operational factors, and public events exactly include this key information. Therefore, information extracted from public events can be used as a priori knowledge, which contributes to the prediction of professional domains. In recent years, many scholars have conducted researches in the professional domains associated with time sequence, their research methods can be divided into: (1) methods based on ARIMA model or regression model [1]; (2) methods based on neural networks or deep neural network model of depth method [2].

Some of these researches combined with prior events, but most of them use the impact factor pre-set by industry experts, but few combined sources of information of public events with business events prediction of professional field of study. Although models based on ARIMA have the advantage of simple network and compatibility for time sequence related prediction tasks, still it is difficult for them to generate satisfying prediction because the models are unable to effectively analysis emergencies and non-regular periodic events or meet the assumptions of stationarity. Hence, these models cannot provide reliable results. Models based on neural network are relatively reliable on exploring non-linear factors, but most of these models used artificial neural networks. The introduction of deep learning is still in research; besides, its adaptability of time sequence is limited.

To solve the above mentioned problem, this paper proposes an LSTM algorithm based on deep neural networks and establishes a LSTM dynamic prediction a priori sequence generation model, which use information obtained from public event sources as a priori information, and use the combination of prior information and data of professional domains as the input of LSTM model to generate sequence [3]. After the training of the model, a priori information of public events sources would be added to every step of the real-time dynamic forecast, together they will predict the result of next time step. This paper combines the LSTM algorithm and public domain data sources to train the model in the field of professional domains prediction tasks, experimental studies show that (1) in addition to traditional natural language processing, songwriter generation and so on, LSTM algorithm can also be adopted to support data prediction through sequence generation. And this model has a certain degree of versatility; (2) combined with a priori information of the public event sources, the model can correctly predict data event trends, which indicates that the model has a better accuracy; and (3) the improved LSTM algorithm can support the prediction of dynamic splicing new data.

2. Data prediction model for professional domains by combing priori information

2.1. Data of public domains events and data information sources of professional domains

Some of the emergencies and non-regular periodic events have been pre-arranged, such as arrangements for competition events, activity reports, broadcast programs, these types of information can be extracted prior to the event on the Internet. Therefore, these events are called prior public events. The reason that the predicted business data in professional domains is unable to fully reflect its regularity is because these events do not occur periodically and even trigger emergencies, but they still can use the a priori knowledge.

Based on these conditions, we divide the factors affecting business data of professional domains into two and extract the information of data sources individually. One is data sources of priori public events, whose data can be collected on the Internet, such as on sites with scheduled events and hot events on Weibo. Another is data sources of professional domain events, whose data are often managed by the data maintenance specialist, such as the operator's traffic flow information, traffic management information—vehicle flow and flow of people, public resources management information—electric power, and water conservation information.

Prior public events often have some of the features, such as:

- predictable: you can know event attributes such as size, duration, scope in advance;
- non-periodic: these events show limited periodically characters, because they happen according to specific arrangements, and sometimes unexpected events occur;
- obtainable: a large-scale of event data, and event attribute information is available on the Internet and easy to be collected.

2.2. Status quo of domestic and abroad researches

Currently most data prediction tasks of professional domains are confined in prediction tools or model algorithms such as probability statistics, gray model, Markov chain, ARIMA model and traditional neural network. And all of them use business data as the only source of data to predict. This paper proposes a model to improve the following aspects (1) most data prediction of professional domains was not combined with external a priori public information; (2) traditional business data prediction algorithms only applied limited time sequence and deep learning; (3) LSTM sequence generation algorithm is not sufficient to support dynamic a prior information splicing. Using LSTM-RNN (short and long term memory recurrent neural network) to build overall prediction model, this paper combines information from two event sources and uses attributes and characteristics of both sources as input to train the model. In the end of this paper, the prediction will be given.

2.3. LSTM dynamic prediction model for a priori sequence generation

First, both a priori public events and professional domain business prediction in this paper are related to time sequence. Time sequence prediction tasks face the following problems: (1) in the time sequence of the events are autocorrelation and non-stationary. It requires the model to adapt to the

IOP Conf. Series: Journal of Physics: Conf. Series **976** (2018) 012007 doi:10.1088/1742-6596/976/1/012007

autocorrelation and partial correlation of time sequence. At the same time, after the introduction of events influence, as it takes a long period of time to change, the attribute of time sequence is non-stationary. The non-stationarity should be set as a prerequisite for model design. (2) the business data is correlated in terms of time transfer, because they all use the information of previous sequence, which results in data memory. Therefore, the model needs to be able to memorize for a period of time and analyze the latest record of the event occurred and the data sequence changes of several previous events.

Second, most of the previous model predictions adopted closed internal data prediction, which is relatively simple and unable to analysis external data. Also, most of them use traditional neural network to predict the future business data distribution. These mathematical statistical models cannot locate the timing of public events. The stationary distribution assumption of ARIMA model does not meet the demand of this scenario. So this paper use neural network RNN-LSTM deep learning model to predict the result.

The working process of conventional LSTM sequence generation model is—establish model through input data training, submit predetermined time step data to the model through using it, thereby after a predetermined time steps, the model can use its parameters and the context provided by previous time steps to successively predict the entire generation sequence and produce the final prediction result. But such predictions model cannot analyze multiple data sources, therefore, they are unable to generate a dynamic prediction or reasonably display the data tendency of emergencies or non-periodic event of professional domains, or effectively improve the prediction accuracy of the model.

Thus, this paper uses the spatial and temporal information as the prerequisite and a priori knowledge of the new model. Deep learning neural network can be adapted to non-linear issues of time sequence. Also, based on the cyclic neural network RNN-LSTM, this model can both solve the problem on a long term memory of time sequence and improve the prediction accuracy by changing the network structure and adding context and a priori information to nerve cells of the network. All these factors will contribute to a more accurate prediction.

3. Introduction of the Technology

3.1. Recurrent Neural Networks

Recursive neural network (shown in Fig. 1) is an artificial neural network for processing timing sequences. It is a cyclic neural network that can be used in tasks such as speech processing, picture description, text generation, text translation, sequence generation and prediction, etc. Because it adopts an error calculation method called Back-Propagation Through Time (BPTT), in which the traditional RNN cannot solve the data association problems that depends on long time. This results in disappearance or the explosion of the gradient. In order to solve the problem, experts including Hochreiter proposed an improved LSTM (long short term memory) in 1997 to resolve the gradient disappearance [4].

As is shown in Figure 2, LSTM proposes a gate mechanism to determine the accumulation of information. It is mainly dealing with problems of sequence of time. Its availability and accuracy in domains including natural language processing and voice processing are proven [5].

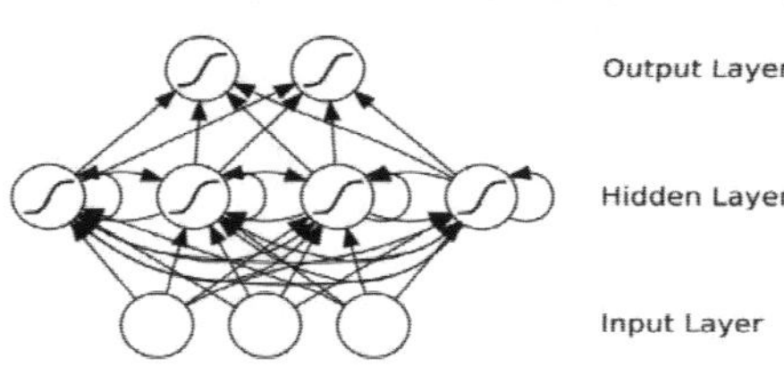

Figure 1 RNN model

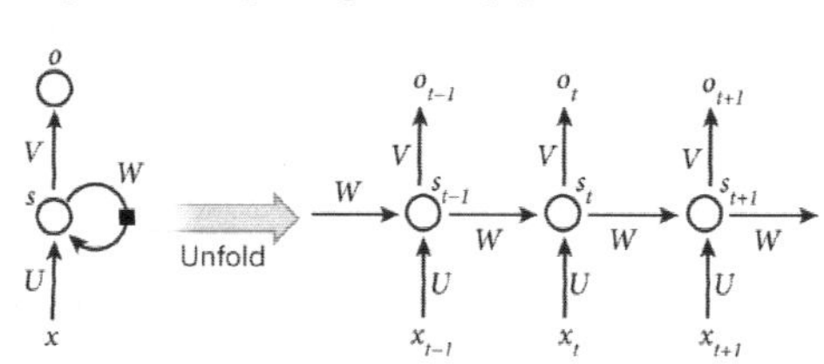

Figure 2 RNN unfold model

LSTM is composed by input gate, forget gate and output gate. As the basic logic units, these gates match the data through weight and the activation function, and control the input, output and forget part of the entire network via the gate. Each neural network module would transfer the message in a cell to the next module. In the following formula, i_t, C_t, f_t and O_t represent the input gate, the state of the cell, forget gate and output gate corresponding to t. LSTM uses the state of cells to record information transferred in neurons. It runs across the neural network with linear interaction. The error function of selective memory feedback is adopted to adjust network weights, forming the predictive model. This allows the long-term message to be passed along the time in multiple recursive neural networks [6].

3.2. *Structure and Principle of LSTM*

Cell state

(1) Forget gate: The forget gate is composed by sigmoid activation function, weight matrix W_f and bias matrix b_f. x_t refers to the input information at the current time step. h_{t-1} refers to the corresponding output of the time step.

$$f_t = \sigma(W_f \cdot [h_{t-1}, x_t] + b_f) \tag{1}$$

(2) Input gate update information: Input gate controls the influx into neurons. i_t refers to the information to be updated of the time step generated by the sigmoid gate. $\widetilde{C}_t$ refers to the retained information of the time step calculated through tanh.

$$i_t = \sigma(W_i \cdot [h_{t-1}, x_t] + b_i) \tag{2}$$

$$\widetilde{C}_t = \tanh(W_c \cdot [h_{t-1}, x_t] + b_c) \tag{3}$$

(3) Updated state of the cell: The equation calculates the cell status. C_t refers to the state of the cell of the present time step. C_{t-1} refers to the one of the previous time step.

$$C_t = f_t * C_{t-1} + i_t * \widetilde{C}_t \tag{4}$$

(4) Output gate: A control gate output is the output of neuron portion. After the layer is obtained using the sigmoid cell state requires the output portion, is through Layer determines which cells need to be output state.

$$O_t = \sigma(W_O[h_{t-1}, x_t] + b_o) \tag{5}$$

$$h_t = o_t * \tanh(C_t) \tag{6}$$

LSTM sequence generation can fully demonstrates the regular change of certain data. It functions the best when the time is fixed and the change is regular. When processing natural language, for example, the sequence generation of the text, after a large amount of training, the model can memorize the language style and language features. Given certain input, it can predict changes in certain time period, and can memorize the features of the provided text, record them in the cell status, observe the features, and provide the corresponding prediction.

4. Data prediction of public events based on improved RNN-LSTM

4.1. *Experiment Design*

(1) Using information source of public events

Information of public events is collectable. One can easily get its properties, such as duration (time), scale, and scope of activities and so on. In this study, relative attributes are duration and weight of the public event influence. These properties are quantified, and combined with influence weights provided by experts. Together they serve as the weight of the public event. Weights are categorized by its volume in specific area, such as concert's weight is heavier than the one of sport events. Information of weights of common events shall be gathered, such as small exhibition venue, events and games. When there is no public event, fixed constants are set, such as zero in usual time, or the peak volume. (The follow-ups of public events are not discussed in this paper).

(2) LSTM sequence generation model

This paper intends to use the predictive function of LSTM sequence generation, and incorporate the source data into data from specific areas in public domain. By designing the training and validation format, data extracted from public events are put into LSTM training mode, improve the verified data source, dynamically assemble public data sources at each time step, so as to provide prediction when the data of the initial time step is given and data of each following time step is incorporated.

4.2. Dataset

4.2.1. *The combination of data of public domain data and professional domain.* In this study, data sets of the volume of base station round Beijing Workers' Stadium are incorporated, with the 4G data volume of Beijing Workers Stadium as the main body of the predictive model. The data sets are divided into a training set and test set.

With specific attributes of a specific event, we can predict the volume of a specific business, such as telephone traffic, electricity flow, traffic flow, visitor flow and so on. These data of specific areas would be influenced by non-periodic events or incidents. Therefore, it requires information of events in public domain.

4.2.2. Pre-processing data

(1) Public event correlation model: Collect attributes in public events, and assign weights to public events according to their scales. Put the corresponding events in the time step in accordance with the event's time information and form new training data set.

(2) Normalization: Process date to (- 1, 1)

(3) Sliding window: Change to the mode of sliding window. The time step is the variation. Pile the training data and test data in the form of sliding windows to achieve a better fit results.

$$x_1, x_2, \dots, x_n = x_1, x_2, \dots, x_n, x_2, x_3, \dots, x_{n+1}, \dots, x_{n+1}, \dots, x_{2n} \ (n = timestep) \quad (7)$$

$$y_1, y_2, \dots, y_n = y_1, y_2, \dots, y_n, y_2, y_3, \dots, y_{n+1}, y_n, y_{n+1}, \dots, y_{2n} \ (n = timestep) \quad (8)$$

4.3. Establish improved LSTM model

This article is based on Tensor flow framework of Python and RNN-LSTM to build the model. It uses sequence to sequence LSTM the network to establish an unidirectional RNN network with LSTM of 4 base unit layers.

(1) Introduction of training mode

When establishing four-layer LSTM model, three to four hidden layer can make it faster to converge, effectively reduce the number of cell units in the hidden layer, and shorten the training time. More hidden layers are more likely to cause the model to overfit.

This study uses four-layer LSTM network as training model, and input both x_e, data of public events, and x_p, data of specific area as the input x(x_e, x_p). The design of four-layer LSTM model is as shown in Figure 3.

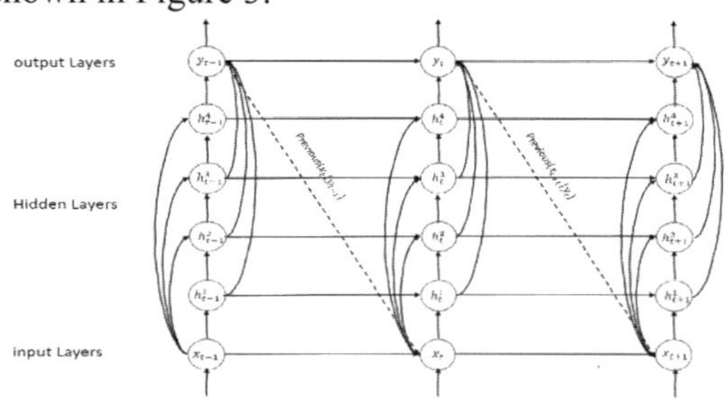

Figure 3 improved LSTM model

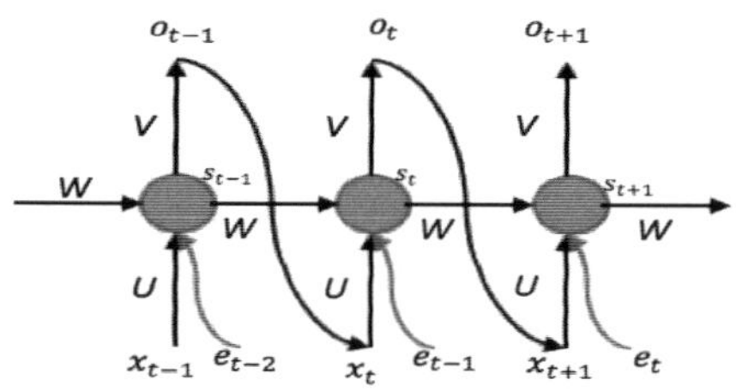

Figure 4 unfold LSTM cell with public event

h_t is the hidden layer node unit of From h_t^1 to h_t^4 is four hidden layers. t is the time step.

A hyperparametric n is required before training. n refers to certain time steps. The sequence observes n steps of input and calculates the true value of the output. The error between the prediction at the end of n steps and the previous batch is calculated. The error is passed back along the timeline and simulate fitting. During training, n steps of input would enter t time steps. Data of each time step t is as shown in equation:

$$y_{timestep} = y_{11}, y_{12}, \cdots, y_{1n} \quad (9)$$

In each batch, each time step t would produce a predictive sequence value y_t. Together with the expected output value of the sliding window $pred$, we can calculate out cost. The model adopts squared error function to calculate error return. cost is the minimized target of the calculation, as shown in the following equation:

$$cost = (pred_t - y_{1t})^2 \quad (10)$$

(2) Improved forecasting series model way

As is shown in Figure 4, the sequence generation method uses the result of a time as the input of following time step. Dynamically input the weight of the public event of next time step as the input value the next prediction.

$$x_t = (y_{t-1}, x_{e_{t-1}}) \quad (11)$$

(3) Testing accuracy:

Testing data: Testing data requires $p_1 \dots p_n$, data of a specific area of a certain time step n, as well as data of events in public domain of the correspondent time. After the iteration of 70 steps, it generates the prediction data of 70-n (n <70) step. As described above, output y of each time step would be the input of the next time step. Together with the dynamic incorporation of data from public events, they will form a new input to predict. And perform the ultimate verification of the model's accuracy of result and tendency.

$$y_t = \text{previous}(\text{tuple}(x_t, x_{e_{t-1}}) | y_{t-1}) \quad (12)$$

5. Experimental results and analysis

This is experiment is based on Tensor Flow depth learning framework and improved algorithm of LSTM in python. Tensor Flow is an artificial intelligence system google released. Compared to CAFFE, it is more adapted to the cycle of neural networks and deep learning scenario of LSTM. Compared to Torch and THEANO ect., it has better adaptability and is more easily to be understood.

5.1. Model training and evaluation

(1) Hyper-parameters

Required experimental initial parameters are: learning rate, number of hidden layers, the number of hidden layer nodes, gradient descent clipping ratio, the iteration number, batch number etc. The experiment maintains a single variable to calculate the error gradient of the sequence adopting forward transmission method of BPTT, and to update the weight values. The experiment has four unit layers, LSTM hidden layers and input layers connected to the front (back) of the hidden layer. When there is no significant reduction of error, the iteration terminates, so that the model converged.

(2) Evaluation

To evaluate the model and the prediction of the model, this paper will apply parts of the public events information, and uses a ratio of the predicted value and the real values to showcase the accuracy of the prediction.

The tendency of the volume of the specific area is illustrated in line chart, in order to show whether the prediction of the tendency is correct when the impact of public events is intense.

5.2. Training results

(1) Business trend data to predict professional domains

As is shown in Figure 5, the predictive model functions well when predicting tendency. It can capture small fluctuations of data of small events, as well as the increasing regions of the change of the data when public events happen. It can also simulate rising/falling speed of the data when it is supposed to change. And when there is no impact from public events, the prediction is very accurate.

Meanwhile, the prediction of peak value under the influence of public events is not accurate enough. But it can provide solid prediction of the information of when the value is getting to the peak. In addition, the prediction of falling region under the influence of public events is not accurate enough.

Due to the error in decreasing region would be passed down with time, the prediction of data after the public event is damaged to some extent.

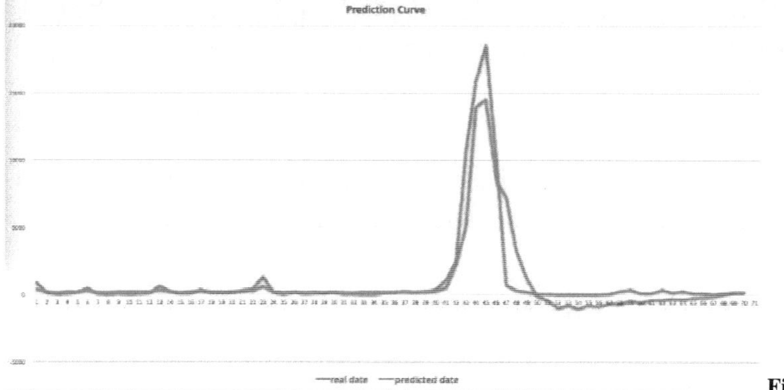

Figure 5 trend of real data and prediction

(2) Model's relative accuracy
As described above, the location of large public events is step 38-50 in the figure. As shown in Figure 6, error is small when there is no public event happening, and the prediction is more accurate. When the prediction for large public event falls in the declining region, the accuracy is not good. In this region, the error is relatively bigger. Since the data has a substantial magnitude, normalization will result in amplification of small error. But the capture of public events and incidents and the prediction in rising region are relatively accurate. Error ratio can be controlled at about 12%. The model may occasionally make a larger error, but it does not affect the subsequent prediction.

Figure 7 shows the error ratio of large-scale public events (ie 38-50 steps). It shows that the prediction after the event is less accurate then the one of the rising region. Accuracy is more guaranteed when the event is rising.

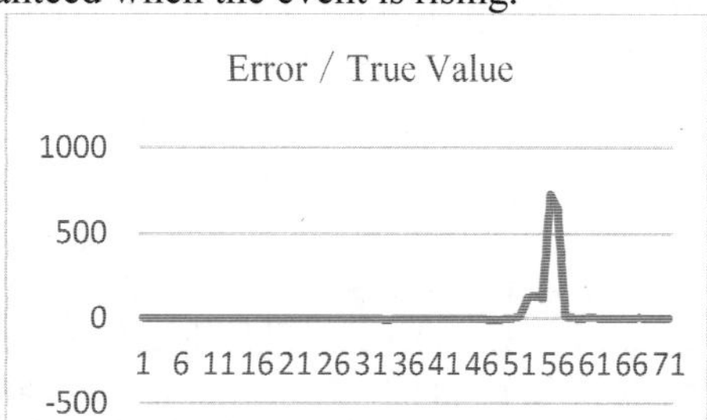

Figure 6 error/true value

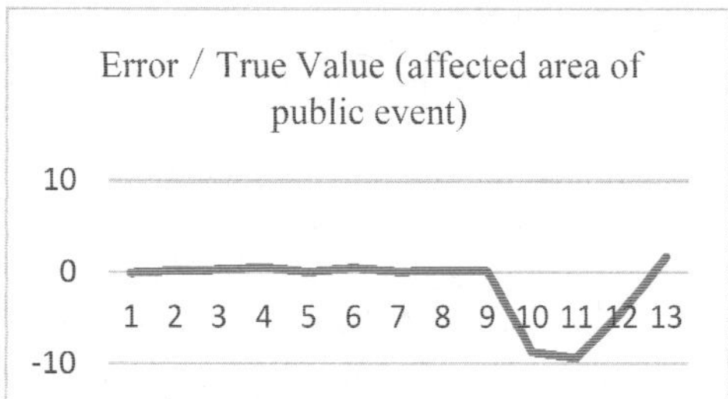

Figure 7 error/true value (affected area of public event)

5.3. Analysis of training results
(1) The number of cells in hidden layers: the size of cells in hidden layer is 140. It proves that the higher number of cells in hidden layers can perform better fit with the data peak.
(2) LSTM layers: the experiment adopts RNNCELL network with 4 layers. The more layers are, the faster convergence of the model is. When there are four layers, the gradient explosion is less common.
(3) Dynamic learning rate: the method does not drop when the anti-conditions, the application rate of dynamic learning occurs in the model accuracy, a small decrease learning rate, this model was observed to a maximum consumption current batch error before two batches of training average error consumption, decreased learning rate of 0.8 times the current, in order to improve the fitting accuracy.
(4) Learning rate: The model's learning rate is 0.007. The higher the learning rate is, the better the fit to the peak value would be; the lower the learning rate is, the better the fit to the valley value would be.
(5) Time step: The prediction is conducted every 24 hours in this study. Since the unit is hour, we adopt 24 hours a day as a period of time, which is a relatively short time for fitting.

(6) Batch: The number of batch is 475. The experiment shows that in each batch of training data, a large change in specific area shall be included to produce more accurate model. The outcome is more ideal when extreme values are included.

Because of the need of predetermined setting of time steps when generating a sequence, which means the prediction can only be performed after these time steps, when the dataset is in sufficient, the usage of sliding windows can largely improve the fitting as it returns the error after the prediction of each time.

Conclusion

The model can correctly predict the tendency of change after public events causes data changes. It can also capture fluctuations caused by small events. The accuracy of the model is consistent. The prediction of the time steps at the peak is less accurate than common, but the prediction of data's tendency is correct. The overall performance of the model in a long sequence generation tasks is in line with expectations. LSTM model can provide relatively accurate prediction by sequence generation when proper information is given.

Error Analysis: Due to the sequence generation method, the error in the previous steps would be accumulated, and passed down to the later prediction. The error is bigger when large-scale public events occur. However, it does not significantly sabotage the prediction.

Improvement: More data of the volume can be input to improve accuracy. Input more influence of public events into the training dataset in order to improve the prediction of peak value and the model of influence.

References

[1] Erasmo Cadenas, Wilfrido Rivera, 2010, Wind speed forecasting in three different regions of Mexico, using a hybrid ARIMA-ANN model. Renewable Energy

[2] Wu JiaoJiao, 2015, The Research of Gas Concentration Prediction Based on Space-time Neural Network Model, China University of Mining and Technology

[3] Alex Graves, 2013, Generating Sequences with Recurrent Neural Networks, computer science

[4] Sepp Hochreiter, 1997, Long Short-Term Memory, Neural computation

[5] Hu XinChen, 2015, Research on semantic relation classification based on LSTM, Harbin Institute of Technology

[6] Kai Sheng Tai, 2015, Richard Socher, Christopher D. Manning, Improved Semantic Representations From Tree-Structured Long Short-Term Memory Networks, computer science

Obstacle Detection using Binocular Stereo Vision in Trajectory Planning for Quadcopter Navigation

Albert Bugayong and Manuel Ramos, Jr

Electrical and Electronics Engineering Institute, University of the Philippines Diliman, Quezon City, Philippines

Email: albert.paolo.bugayong@eee.upd.edu.ph, manuel@eee.upd.edu.ph

Abstract. Quadcopters are one of the most versatile unmanned aerial vehicles due to its vertical take-off and landing as well as hovering capabilities. This research uses the Sum of Absolute Differences (SAD) block matching algorithm for stereo vision. A complementary filter was used in sensor fusion to combine obtained quadcopter orientation data from the accelerometer and the gyroscope. PID control was implemented for the motor control and VFH+ algorithm was implemented for trajectory planning. Results show that the quadcopter was able to consistently actuate itself in the roll, yaw and z-axis during obstacle avoidance but was however found to be inconsistent in the pitch axis during forward and backward maneuvers due to the significant noise present in the pitch axis angle outputs compared to the roll and yaw axes.

1. Introduction

Mobile robotics deals with the creation of robots capable of locomotion through land, water and air. It also deals with intelligent navigation using data obtained from sensors for self-localization, mapping and path planning [1], [2].

Sonars, IR sensors, laser range finders, and multiple cameras have been used in mobile robots to obtain distance information from its environment for obstacle avoidance and navigation. Sonars, IR sensors and laser range finders, however, have the disadvantage of obtaining information only at a specific point in space. As a result, they are used in large quantities to form sensor matrices or are mounted onto motors to perform a sensor sweep on the surroundings to obtain more information from the environment. Some researches, notably from Saxena, et.al, used monocular vision and structured lighting to obtain distance information from its environment [3]. Stereo vision is ideal as it is passive and is less computationally expensive compared to monocular vision.

This project focuses on indoor navigation for reconnaissance in closed environments such as buildings and mining areas.

2. Methodology

2.1. Sensor Fusion

Gyroscopes are susceptible to drifts or accumulations of angle measurement errors while accelerometers are prone to high frequency noise due to vibrations. A high-pass filter is applied to the gyroscope to cancel out drift while a low-pass filter is applied to the accelerometer to minimize noise from vibrations [4]. A complementary filter was used to fuse gyroscope and accelerometer data to obtain accurate pitch yaw and roll angle estimations. The equation for the complementary filter is given by:

$$\theta = (\alpha) * \left(\theta + \dot{\theta}_{gyro_{reading}} \cdot dt\right) + (\alpha - 1) * \left(\theta_{accel_{reading}}\right) \quad (1)$$

where dt is the sampling period and α and is defined by:

$$\alpha = \frac{\tau}{\tau + t_s} \quad (2)$$

where τ is the time constant which determines both the high pass and the low pass filter cut off frequency and t_s is the sampling period [4], [5].

2.2. Stereo Vision

For cost matching, the sum of absolute difference (SAD) is the most commonly used matching criteria due to its low complexity and good performance [6]. The equation is given by:

$$SAD = \sum_{(i,j)\in W(x,y)} |IL_{(i,j)} - IR_{(i,j+d)}| \quad (3)$$

where I_L is the left pixel intensity, I_R is the right pixel image intensity, W is a square window surrounding pixel (x, y) and d is the disparity value [6].

The sum of absolute difference takes the absolute difference between the reference window from the first image and the target window in the second image. The reference window is moved pixel by pixel across all target windows until the associated cost is minimized [6]. The depth of an object is measured from the baseline between the camera centers. The disparity of an object is the difference of its projections from the two images [7]. The depth can be computed using the equation:

$$Z = \frac{bf}{d} \quad (4)$$

where Z is the depth calculated, b is the baseline or the distance between the two cameras, f is the camera focal length and d is the disparity [7].

A vector slice was obtained from the center row of the disparity map. The depth from the camera's baseline for each disparity was computed using equation 7. This distance was then used to obtain obstacle locations from the environment and served as a basis for the execution of a proper obstacle avoidance maneuver.

2.3. Motor Control

A PID controller was applied separately to the Altitude, Pitch, Yaw and Roll controllers [8], [9].

$$PID_\theta = k_p(\theta_{error}) + k_i(\theta_{error,accumulated}) + k_d(\theta_{error,predicted}) \quad (5)$$

where $\theta_{error} = \theta_{current} - \theta_{desired}$ [8].

The pitch, roll and yaw and altitude each have their own set of PID controllers. Each PID controller is represented by equation:

$$duty_n = PID_{pitch} \pm PID_{roll} \pm PID_{yaw} \pm PID_z \pm duty_{base,throttle} \quad (6)$$

where $duty_n$ is the input PWM signal to the ESC for each motor n, $duty_{base,throttle}$ is the base duty or constant throttle required for quadcopter elevation [5], [9].

2.4. Trajectory Planning

To construct the Cartesian Histogram Grid, a 1D depth vector slice from the 2D disparity map is obtained. The VFH algorithm [2], [10], [11] was modified to suit the scope of this research. Due to the absence of position tracking, once the desired heading is obtained, the current map information is disposed of in preparation for the next sampling. The first stage of this process is to create the occupancy grid. 5 meters was the set maximum distance wherein obstacles are detected. Consequently, the grid dimensions were set to 5 by 5 meters and divided into 1089 cells, equivalently, 33 by 33 cells. The resolution of each cell is effectively 15 centimeters. The cartesian coordinates for each occupied cell were then obtained through the following equations [12]:

$$x = zcos(\theta) \quad (7)$$

$$y = zsin(\theta) \quad (8)$$

where x and y are the coordinates of an occupied cell, z is the depth from the depth vector, and θ is the depth's corresponding angle computed by:

$$\theta = pixel(i) \times \left(\frac{fov}{nop}\right) \quad (9)$$

where *i* is the pixel row coordinate, *fov* is the field of view and *nop* is the number of pixels in the row.

For the second stage of this process, the obstacle magnitude vectors $m_{i,j}$ and direction vectors $\beta_{i,j}$ were computed for each cell obstacle using equations 12 and 13.

$$m_{i,j} = (c_{i,j})(d_{max} - d_{i,j}) \quad (10)$$

$$\beta_{i,j} = \tan^{-1}\left(\frac{y_i - y_o}{x_i - x_o}\right) \quad (11)$$

where $c_{i,j}$ is the certainty value set to 1 assuming that the depth obtained is the same as the actual environment, $d_{i,j}$ is the distance of the robot from the cell in centimeters and d_{max} is the maximum distance possible which is 5 meters or 500 centimeters. The obstacles were then grouped into sectors to form the primary polar histogram using equation 14 to 16 where α =5° is the angular sector resolution and k is the discrete quantized multiples of α *and* H^p is the polar obstacle density (POD).

$$k = INT\left(\frac{\beta_{i,j}}{\alpha}\right) \quad (12)$$

$$\alpha = \frac{360}{n} \quad (13)$$

$$H_k^p = \sum_{i,j\in C_\alpha} m_{i,j} \quad (14)$$

The sectors were labeled in the opposite order as the map generated was a mirror image of the actual environment. For the third stage, a binary polar histogram H^b was derived from the primary polar histogram by filtering the POD's through a threshold was experimentally obtained.

$$H_k^b = 1, \text{ if } H_k^p > \tau$$

$$H_k^b = 0, \text{ otherwise}$$

At the fourth stage, the candidate valleys were obtained from the resulting free sectors from the binary polar histogram. The valleys were then labeled as free or blocked if too narrow. At the last stage, the center sector location of each candidate valley was obtained and the valley selected is the valley closest to the center or current heading of the quadcopter.

3. Hardware Implementation

The Raspberry Pi Compute Module with IO Board v.1 was selected as the single board computer (SBC) for this project as it comes with two Camera Serial Interface (CSI) ports. This was essential to enable the use of a stereo camera pair [13]. The Raspberry Pi Model B+ was used for motor control and trajectory planning blocks. Two lightweight Raspberry Pi camera modules were used for implementing stereo vision [14]. OpenCV, also known as Open Source Computer Vision Library, is an open source image processing software library used in implementing Stereo Vision in this research [15]. Adafruit's 10 DOF IMU containing a L3GD20H gyroscope, LSM303 compass and the LSM303 accelerometer in a single breakout board was used for this research. A HC-SR04 ultrasonic sensor was used for altitude estimation. The DJI F450 quadcopter frame used has a range between 600g - 1200g. The total weight of all components was found to be 955.5g ensuring the payload of the quadcopter was within its limits.

4. Results and Analysis

4.1. Stereo Vision

It can be seen on figure 1 that the errors in distance measurement from the stereo vision sensor were 20 cm to 50 cm away from the actual distance which is sufficient for the application in this project. The discontinuity errors in the depth map slice was taken into account in the VFH block of the system by marking them as blocked during the sector clustering stage.

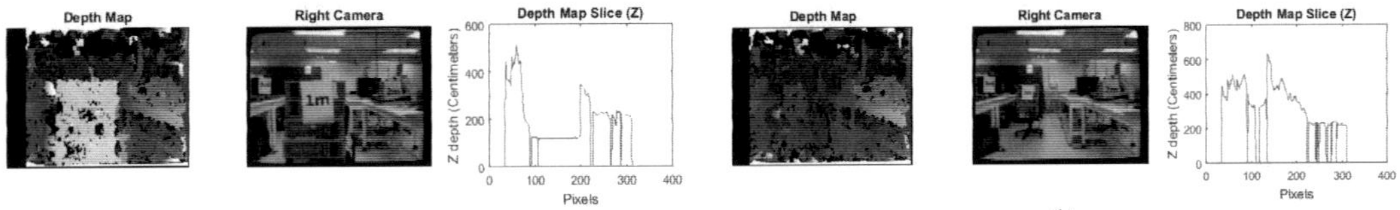

(a) (b)

Figure 1: Depth slice with subject at 1 meter (a) and 3 meters (b) from the baseline.

4.2. Motor Control

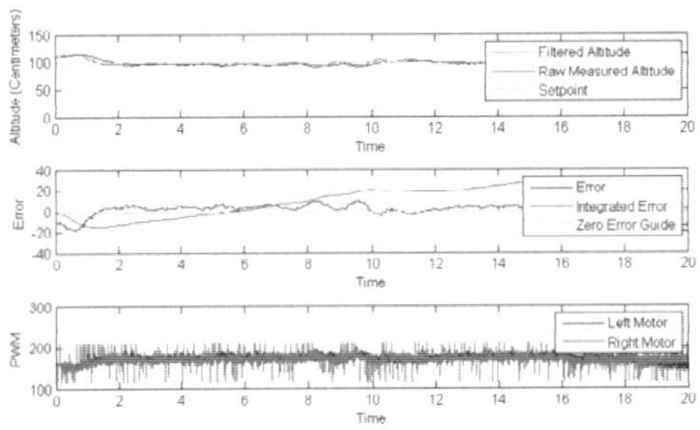

Figure 2: Altitude control flight response at 100cm setpoint

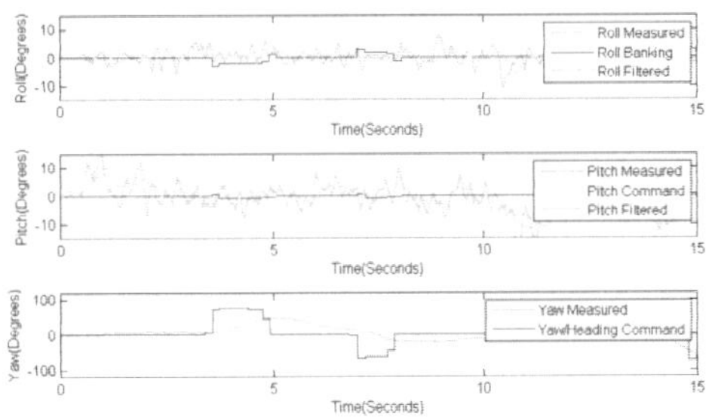

Figure 3: Trajectory Planning – Command vs Response – Left Entry

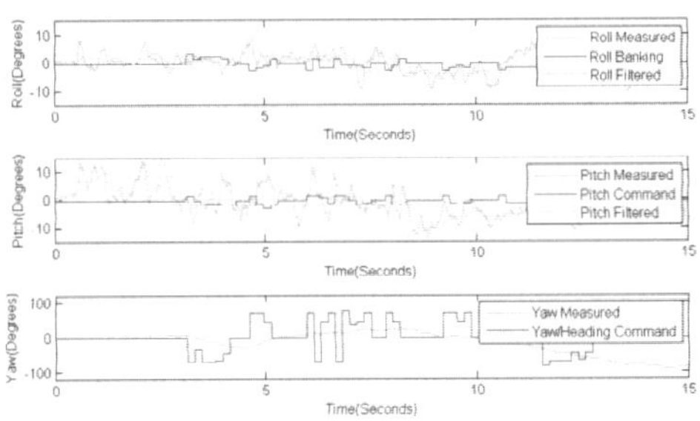

Figure 4: Trajectory Planning – Command vs Response – Right Entry

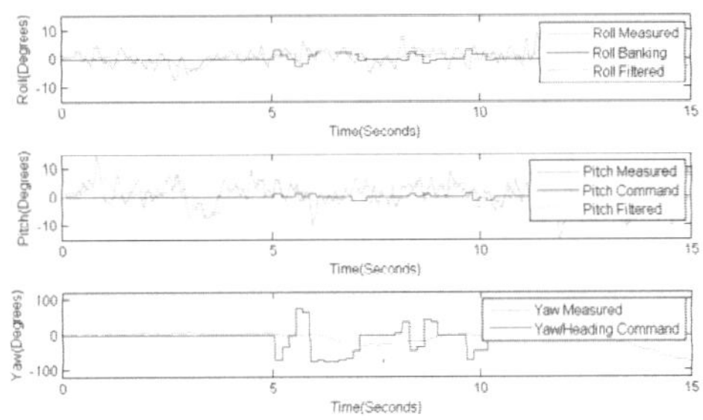

Figure 5: Trajectory Planning – Command vs Response – Reverse Maneuver

4.2.1. PID Controller. The Single Loop Altitude PID was used for altitude control. It could be seen at figure 2, that the setpoint of 100cm during hover was reached with minimal to no oscillations.

The cascaded PID controller consists of two loops. The inner control loop or the rate loop controls the angular rate while the outer loop or the stabilize loop controls the angle. A cascaded PID controller was used to obtain angular velocity data from the gyroscope which was not affected by high frequency vibrations to reduce noise entering the system from the D controller. The resulting cascade is a P-PID control loop. The current refresh rate of the control loop was 500Hz to 540Hz.

4.2.2. Trajectory Planning

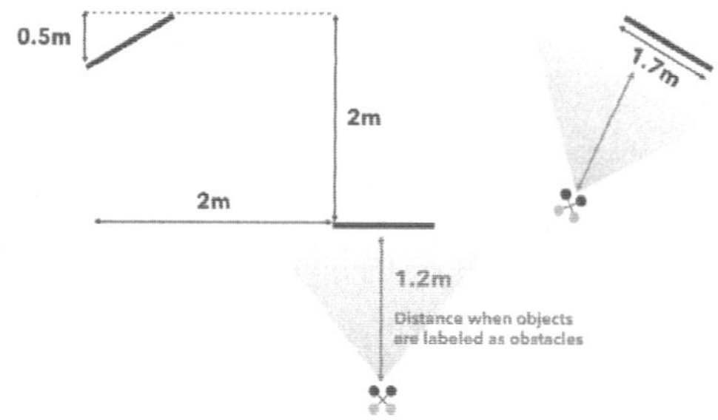

Figure 6: Obstacle Set-up

In the tests conducted using the set-up in figure 6, consistent correct steering along the yaw axis inflight was observed given the appropriate velocity was attained when approaching the obstacle. A fast approach caused the obstacle not to be detected because of a slow frame rate of 6hz. This framerate entailed a delay of 16ms in average thus an ideal traveling velocity was required to compensate. However, when an obstacle was detected during a fast approach, the quadcopter might be too near causing oscillations in the steering decision requiring a reverse maneuver to be performed. The oscillation was caused by the low sector resolution set at 5° which was equivalent to only 12 sectors. This was a seldom problem given the correct velocity was maintained. It can also be noted that if too many commands on each axis overlap such as when the quadcopter was oscillating between two obstacles, sudden jumps in altitude were observed.

In the data observed, the pitch and roll axis readings from the complementary filter were very noisy with ±3° in the roll axis and ±8° error in the pitch axis when hovering due to motor vibrations which caused the readings to be unreliable with commands less than the magnitude of the noted errors. However, it was found that using commands greater than 5° on the pitch axis caused fast approaches and doing the same with the roll axis caused severe banking leading to overshoot, at worse cases overpowering the pitch command impeding forward movement. Thus, a lower range of commands was selected, ±3° for both the pitch and the roll. Despite the problem with the readings, there was a visible response in the pitch and roll axes while given a command inflight. This was most apparent with the roll axis which was enabled consistent banking. The default pitch command for slow forward travel however was not consistent.

4.2.2.1.Left Entry. During an entry attempt at the left side of the set-up, Figure 3 details the response of the quadcopter in each axis given a command. It could be observed that due to the noise in the pitch and roll axes, it was difficult to attain the roll and, more so, the pitch axis setpoints. The yaw, however, was less prone to noise influence due to the large angle value setpoints which was much greater than the noise from the complementary filter output. The yaw was able to steer correctly, albeit, with a delay approximately 1.5 seconds. The PID for the yaw axis was tuned for this smooth slow response as a fast response would cause the quadcopter to jump along the z-axis. The first yaw command issued was positive to turn the quadcopter counter clockwise together with a pitch command of 0.6° to slow down and -0.7 to gain speed, and a roll command of -3° and -2° to bank left. As the last positive yaw command was issued, the roll performs a counter bank to avoid drifting along the horizontal and the pitch axis resumes with the default pitch command as it approaches the next obstacle. The second yaw command was negative to turn the quadcopter to clockwise as it approaches the left obstacle and performs the same pitch maneuvers as described before but performing the reverse of the previous banking maneuvers.

4.2.2.2. Right Entry. Figure 4 details the response of the quadcopter in each axis as an entry in the right side of the set-up was performed. It can be observed that due to the noise in the pitch and roll axes, it was difficult to attain the pitch and the roll axis setpoints. To avoid the first obstacle, the quadcopter this time turned clockwise with a negative angle on the yaw axis, and banked to the right as seen by the positive roll angle commands between 3 and 4 seconds. Simultaneously, the quadcopter pitched upward to slow down as shown in the present positive angle command and then proceeds to decrease its banking angle as it avoids the obstacle. The opposite yaw and roll maneuvers were executed at the 5-second mark as the quadcopter approaches the second obstacle at the right side of the set-up. The same pitch maneuvers were executed simultaneously. The commands after the 6-second mark were to be disregarded as they were data collected as the quadcopter was stowed in preparation for the next trials.

4.2.2.3.Reverse. Figure 5 shows a case wherein oscillation in steering was observed when the quadcopter approaches too fast thus performing the decision making at a distance too near the obstacle. The quadcopter was set to pitch back whenever it encounters a large change in yaw angle commands, a change greater than 45°. The pitch was set to 1.25°. Pulses of this magnitude could be seen between the 5- to 7-second mark in the plot of the pitch axis. This eventually caused the quadcopter to pitch back to gain distance from the obstacle and then attempts again to avoid the obstacle heading towards the right side.

5. Conclusion

The data obtained from the stereo vision sensor was reasonably accurate up to 5 meters. However, the current SBC used in this research has proven to be insufficient to get a standard frame rate of 30 frames per second for a sufficiently fast update rate to provide more current information for the navigation block.

The Raspberry Pi was sufficient for motor control due as it was able to operate the PID Control loop at a rate of 500Hz to 540Hz resulting to sufficient corrections to maintain stable flight. The system, however, requires a fixed velocity when travelling forward to compensate for the delay in obtaining depth information due to the slow frame rate as well as the slow yaw response which was necessary to avoid sudden jumps along the z-axis when turning. Take-off and landing were not implemented requiring the quadcopter to be released at a height above the setpoint during operation. There exist lateral drifts on the X and Y axes with the X axis minimal while the Y axis has a tendency to pitch forward due to the inherent bias of the quadcopter. Inconsistencies in maintaining travelling velocity can be attributed to the significant noise present in the pitch axis angle outputs compared to the roll and yaw axes.

References

[1] G. Cook, Mobile robots: navigation, control and remote sensing. John Wiley & Sons, 2011.
[2] Lounell B. Gueta and Manuel C. Ramos, Jr. Mobile robot navigation using binocular stereo vision MS thesis, University of the Philippines-Diliman, 2005.
[3] Saxena, S. H. Chung, and A. Y. Ng, "3-d depth reconstruction from a single still image," International journal of computer vision, vol.76, no.1, pp.53–69, 2008.
[4] S. Colton and F. Mentor, "The balance filter," Presentation, Massachusetts Institute of Technology, 2007.
[5] Carissa Norielle L. Cruz, Motor Control using a Smartphone for Fixed Path Quadcopter Flight, Undergraduate Student Project, Univesity of the Philippines-Diliman, June 2015.
[6] R. H. Thaher and Z. K. Hussein, "Stereo vision distance estimation employing sad with canny edge detector," International Journal of Computer Applications, vol. 107, no. 3, 2014.
[7] N. Oon-Ee, V. Ganapathy, and S. Ponnambalam, "Modular stereo vision: Model and implementation," Depth Map and 3D Imaging Applications: Algorithms and Technologies: Algorithms and Technologies, p. 245, 2011.

IOP Conf. Series: Journal of Physics: Conf. Series **976** (2018) 012008 doi:10.1088/1742-6596/976/1/012008

[8] Zul Azfar, A.; Hazry, D., "A simple approach on implementing IMU sensor fusion in PID controller for stabilizing quadrotor flight control," in Signal Processing and its Applications (CSPA), 2011 IEEE 7th International Colloquium on , vol., no., pp.28-32, 4-6 March 2011
[9] Bouabdallah, S. "Design and control of quadrotors with application to autonomous flying." PhD diss., École Polytechnique federale de Lausanne, 2007.
[10] Borenstein, J.; Koren, Y., "The vector field histogram-fast obstacle avoidance for mobile robots," in Robotics and Automation, IEEE Transactions on , vol.7, no.3, pp.278-288, Jun 1991
[11] Ulrich, I.; Borenstein, J., "VFH+: reliable obstacle avoidance for fast mobile robots," in Robotics and Automation, 1998. Proceedings. 1998 IEEE International Conference on , vol.2, no., pp.1572-1577 vol.2, 16-20 May 1998
[12] A. Guevara, “Using The Sensor Kinect For Landmark Recognition At Escuela De Ingenier´ıa DeAntioquia”, Escuela De Ingenier´ıa DeAntioquia, Undergraduate Project, 2011
[13] Raspberry Pi Foundation (2014, April 7). Raspberry Pi CMIO [Online]. Available: https://www.raspberrypi.org/blog/tag/compute-module-io-board/
[14] Raspberry Pi Foundation (2013, May 14). Raspberry Pi Camera [Online]. Available: https://www.raspberrypi.org/documentation/hardware/camera/
[15] opencv.org

IOP Conf. Series: Journal of Physics: Conf. Series **976** (2018) 012009 doi:10.1088/1742-6596/976/1/012009

Intelligent Planning for Laser Refractive Surgeries

Wei Wang[1,2,4], Yong Yue[1], Ahmed Elsheikh[2] and Fangjun Bao[3]

[1] Department of Computer Science and Software Engineering, Xi'an Jiaotong-Liverpool University, Suzhou 215123, P.R. China
[2] School of Engineering, University of Liverpool, The Quadrangle, Brownlow Hill, Liverpool L69 3GH, UK
[3] School of Optometry and Ophthalmology and Eye Hospital, Wenzhou Medical University, Wenzhou 325027, P.R. China
[4] Institute of Complex Networks and Visualisations, Shijiazhuang Tiedao University, Shijiazhuang 050043, P. R. China

wei.wang02@xjtlu.edu.cn

Abstract. Refractive error is one of leading ophthalmic diseases for both genders all over the world. Laser refractive correction surgery, e.g., laser in-situ keratomileusis (LASIK), has been commonly used worldwide. The prediction of surgical parameters, e.g., corneal ablation depth, depends on the doctor's experience, theoretical formula and surgery reference manual in the preoperative diagnosis. The error of prediction may present a potential surgical risk and complication. Being aware of the surgery parameters is important because these can be used to estimate a patient's post-operative visual quality and help the surgeon plan a suitable treatment. Therefore, in this paper we discuss data mining techniques that can be utilized for the prediction of laser refractive correction surgery parameters. It can provide the surgeon with a reference for possible surgical parameters and outcomes of the patient before the laser refractive correction surgery.

1. Introduction

In the last two decades, the globe prevalence of refractive errors has risen dramatically. In China, the prevalence of myopia and high myopia is more than the world average, and is associated with education [1~4]. This status will affect the Chinese population quality. Considerable research has been conducted on refractive surgeries and there are publications about the success of laser refractive correction surgery, e.g., LASIK [5~10]. While the patient satisfaction rate of LASIK surgery is more than 90%, it may not be recommended for everybody. There are two main reasons: (1) potentially no significant visual acuity improvement for certain types of patients, and (2) the complications after the surgery [5, 10]. At present, the surgeon performs the surgery prediction with patient's biological features, surgery parameters, theoretical formulas and hypotheses. This prediction is a theoretical result based on which, the surgeon can rely on their experience and surgical equipment manuals to roughly estimate the surgery outcomes [11, 12]. If there is a method to predict surgery parameters without theoretical formulas, hypotheses and the experience of the surgeon, it will provide another reference for the surgeon to adjust the surgery parameters for better surgery outcomes, and the patient could know the possible improvement of visual quality after the LASIK surgery.

The medical history of surgeries might hide the relationships between the pre- and post-operative parameters. This research explores ways to achieve accurate surgical parameters with data mining techniques, and has obtained regression equations of surgical parameters with historical medical data. The goodness of fit, which is used to evaluate regression equations, is more than 0.8, indicating a high accuracy.

2. Related work

Data mining methods are divided into two types, predictive methods (e.g., classification, regression, time series analysis and prediction), and descriptive methods (e.g., clustering, association rule mining, and sequence analysis). Regression is widely used in medicine as a method of predicting, comparing, and evaluating medical procedures and outcomes.

Before laser refractive surgery, Jester *et al.* [13] and Salz *et al.* [14] used fresh human cadaver eyes to analyse the relationship among incision depth, corneal curvature, corneal thickness, corneal diameter and incision length with stepwise regression in radial keratotomy, and introduced a prediction equation of change in corneal curvature correlation after radial keratotomy. While impossible to track biological features and surgery outcomes for the cadaver eyes in the long term, they presented a way to predict refractive surgery outcomes with linear regression. Block and Block [15] used multiple regression to determine the surface of the cornea in radial keratotomy. They only introduced the algorithm and ellipsoid equation, and did not give any solution to the equation. Until 2011, Bao *et al.* [16] and partners solved the ellipsoid equation with multiple regression for 112 eyes. They also figured out that the axial length was the main morphological parameter related to myopia.

Since the laser refractive surgery was invented in early1990s, there has been research on the analysis of historical medical data and the discovery of relationship among patient biological features, surgical parameters and outcomes. For the analysis of patient biological features, Kohlhaas *et al.* [17] found that there were slight correlations among intraocular pressure (IOP), central corneal thickness (CCT) and corneal curvature with regression after corneal refractive surgery. With linear regression for 56 eyes, Bühren *et al.* [18] claimed that the postoperative wavefront error had limited influence on the subjective quality of vision. With linear regression for 706 eyes, Recep *et al.* [19] claimed that the decrease in IOP was related to the decrease in corneal stromal thickness after LASIK. Chen *et al.* [20] and Hamilton *et al.* [21] introduced the correlation between corneal biomechanical properties (e.g., corneal hysteresis and corneal resistance factor), and attempted refractive correction in LASIK. Chang and Stulting [22] presented a method to predict the change of IOP after LASIK with refractive change. Wangsupadilok *et al.* [23] presented a regression equation of IOP and CCT.

For the analysis of surgery parameters, Yi W-M and Joo C-K [24] introduced a linear regression method to evaluate the correlation between corneal flap thickness and preoperative corneal thickness with 69 eyes. Huang *et al.* [25] introduced a method to estimate the deviation in myopia and astigmatism mixed refractive surgery with multiple regression for 523 eyes. He found that the spherical ablation by LASIK had a refractive change 19% greater than by PRK. With linear regression for 102 eyes, Durairaj *et al.* [26] presented the regression equation of actual stromal ablation. Ogasawara and Onodera [27] analysed the correlation of residual stromal bed thickness and regression of myopia after LASIK with linear regression. Choudhri *et al.* [28] found that the corneal thickness was most significantly correlated with the corneal flap thickness with regression. Liyanage *et al.* [29] and Allan *et al.* [30] developed a systematic method for quantifying pre-treatment adjustments for LASIK patients with multiple linear regression.

For the analysis of surgery outcomes, Dutt *et al.* [31] used one-year results of excimer laser photorefractive keratectomy (PRK) to analyse the correlation between the attempted dioptric correction and the achieved correction with linear regression. With linear regression for 50 eyes, Lee [32] discovered that glare and halo symptoms were correlated with the attempted correction of spherical equivalent and astigmatism, but not with the pupil size. In contrast, with linear regression for 92 eyes, Helgesen *et al.* [33] claimed the large pupil size is associated with postoperative visual disturbances during scotopic conditions. For these small size datasets, the results may be disturbed by the outliers in the dataset. With linear regression for 20 eyes, Srivannaboon [34] claimed that the corneal power change was correlated highly with the manifest refractive change produced by LASIK. With linear regression for 43 eyes, Lackerbauer *et al.* [35] claimed that there was limited correlation between the corneal ablation and refractive outcomes for LASIK. Alio *et al.* [36] evaluated PRK outcomes and introduced a predictive model for the refractive changes in the long term with linear regression for 33 eyes.

The related work provided an efficient way to plan and valuate the laser refractive surgery using data mining techniques. This research explores how to predict the corneal maximum ablation depth with data mining. In additional, more details of regression (e.g., standard error and correlation which have been neglected in previous research) will be used to evaluate the accuracy of model and results.

3. Materials and Methods

In this research, the surgery data was collected from the Eye Hospital of Wenzhou Medical University (Zhejiang Eye Hospital). The dataset has 30 items for 1559 eyes of 786 patients in total. There are 4 items of patient demography, age, gender, preoperative examination date and surgery data. There are 6 items of preoperative examination, diopter of spherical power (SP) and cylindrical power (CP), astigmatism axial, best corrected visual acuity (BCVA), CCT and IOP with non-contact tonometer (NCT). There are 5 items of surgery parameters, optical zone dimeter (OpD, in mm), cutting zone diameter (in mm), corneal flap thickness (in μm), corneal residual thickness (in μm) and corneal maximum ablation depth (CMAD, in μm). The data is from three post-surgery examinations, diopter of spherical power and cylindrical, astigmatism axial, BCVA and IOP with NCT, 1 week, 1 month and 3 months, respectively. The data contains information for 786 patients in the age range of 16 to 51. 419 patients are females, and the rests are males. The histogram of preoperative SP is shown in Figure 1. 209 eyes are in the spherical power range of 0 D to -3.00 D. 825 eyes are in the spherical power range of -3.00 D to -6.00 D. 343 eyes are less than -6.00 D. 182 eyes are less than -8.00 D. The histogram of CMAD is shown in Figure 2. Both of these two parameters are with a similar distribution.

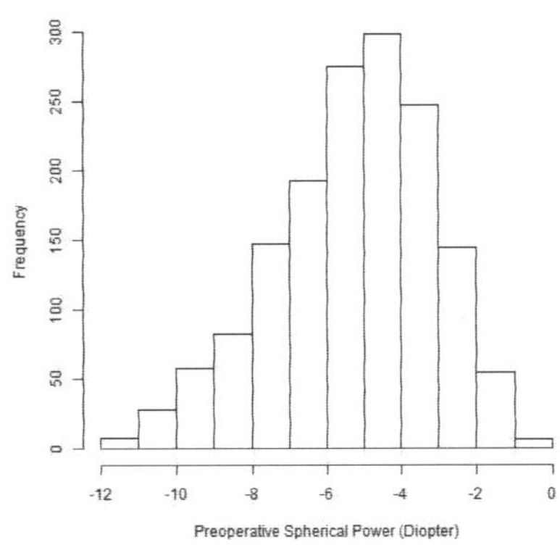

Figure 1. Histogram of preoperative SP

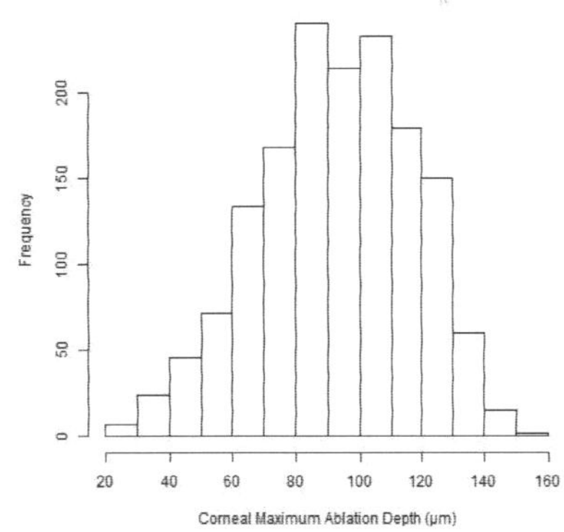

Figure 2. Histogram of CMAD

Table 1. Mean values of all measures

Parameter	Mean Difference±Standard Deviation
Age (year)	23.85±5.90
Preoperative SP (diopter)	-5.15±2.11
Preoperative CP (diopter)	-0.81±0.69
Preoperative astigmatism axis (degree)	86.93±73.63
Preoperative CCT (μm)	540.81±30.24
Preoperative NCT (mmHg)	15.59±2.91
Preoperative BCVA	1.06±0.12
OpD (mm)	6.58±0.42

CMAD (μm)	93.38±24.15
Postoperative SP (diopter)	0.35±0.46
Postoperative CP (diopter)	-0.35±0.34

For the prediction of CMAD, the biological features and optical zone diameter are considered as the known conditions of the regression equations. The mean values of these measures are shown in Table 1. The prediction model of CMAD is studied by correlation and regression analysis. The QR decomposition algorithm is used to complete the regression linear fit in R. A P-value of less than 0.05 is statistically significant.

4. Results

To judge whether the ablation depth and other parameters really have a high correlation, the correlation is calculated between ablation depth and other parameters with a Pearson correlation coefficient, as shown in Tables 2 and 3.

Table 2. Correlation between ablation depth and other parameters

	Age	Gender	Preoperative SP	Preoperative CP	Preoperative axis
Correlation	0.05	0.07	-0.85	-0.35	0.03

Table 3. Correlation between ablation depth and other parameters

	Preoperative CCT	Preoperative NCT	Preoperative BCVA	OpD	Postoperative SP	Postoperative CP
Correlation	0.05	0.07	-0.17	-0.53	0.19	-0.19

According to the correlation among the parameters, preoperative spherical power, preoperative cylindrical power and optical zone diameter have high correlation with corneal maximum ablation depth. The three parameters are used in the multiple regression. The correlation of the ablation depth and the preoperative spherical power is the highest than others; therefore, the preoperative spherical power will be used in single regression, separately.

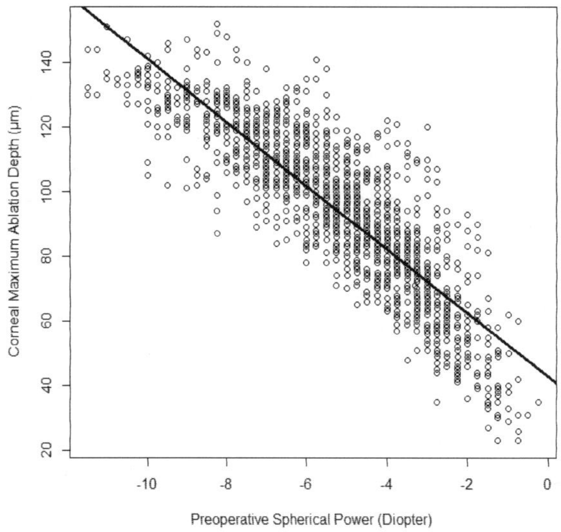

Figure 3. Linear regression between preoperative SP and CMAD

For the preoperative SP and CMAD regression, as shown in Figure 3, the regression equation is shown below. The residual standard error (σ) is 12.36μm. The R-squared is 0.738. The P-value is less than 2.2e-16.

$$Y = 42.7577 - 9.8297x \tag{1}$$

where *Y is expected ablation depth. x is preoperative SP.*

For the multiple regression, the regression equation is shown below. The residual standard error (σ) is 8.35μm. The R-squared is 0.8806. The P-value is less than 2.2e-16. The multiple regression is with less residual standard error and greater R-squared values than single linear regression.

$$Y = -146.0074 - 13.1426x_1 - 11.9439x_2 + 24.5865x_3 \quad (2)$$

where *Y is expected ablation depth.* x_1 *is preoperative SP.* x_2 *is preoperative CP.* x_3 *is OpD.*

To evaluate the efficiency of the regression equation, the predictions are compared between the regression equation and Munnerlyn approximate formula [37].

$$\text{Approximate Ablation Depth} \cong \frac{OZ^2}{3}D \quad (3)$$

where *OZ is optical zone diameter. D is the correction in dioptres.*

The mean-squared-error (MSE) is a measure of the quality of an estimator [38]. The root-mean-square-error (RMSE) is a measure of accuracy, to compare forecasting errors of different models for a particular dataset and not between datasets [39]. For the prediction model with regression, MSE and RMSE can be used to evaluate the accuracy. The comparison includes the mean difference, MSE and RMSE, and the results are shown in Table 4. The multiple regression equation with less MSE and RMSE means the accuracy is better than the accuracy of Munnerlyn approximate formula.

Table 4. Comparison of multiple regression and Munnerlyn approximate formula

	Mean Difference	MSE	RMSE
Multiple Regression Equation	0.01016±6.104959	69.49518	8.336161
Munnerlyn Approximate Formula	21.5491±8.148693	584.3581	24.1735

5. Conclusion

The research has achieved a data-driven model automated method to obtain a possible surgery parameter, the corneal maximum ablation depth. The regression equation is more accurate than the theoretical formula, and more convenient to use than the surgery system. It can also easily estimate the residual stromal depth, which is correlated with a series of postoperative complications.

In addition, more regression models (e.g., gradient boosting regression, Bayesian regression, ridge regression, lasso regression, neural network regression) will be used for the prediction of surgical parameters and outcomes. A comparison among these methods will also be made in the future.

Acknowledgement

This work is carried out in the Department of Computer Science and Software Engineering at Xi'an Jiaotong-Liverpool University (XJTLU). It is partially supported by a XJTLU Research Development Fund project under Grant No. RDF-14-02-34 and the Suzhou Virtual Reality Lab Platform project under Grant No. RRSP1012017029.

References

[1] Xiao O, Morgan IG, Ellwein LB and He M 2015 *Ophthalmology* **122** 1924-31
[2] Rudnicka AR, Kapetanakis VV, Wathern AK, Logan NS, Gilmartin B, Whincup PH, *et al.* 2016 *British Journal of Ophthalmology* **100** 882-90
[3] Li Z, Xu K, Wu S, Lv J, Jin D, Song Z, *et al.* 2014 *Clinical & Experimental Ophthalmology* **42** 379-84
[4] Kuang TM, Tsai SY, Liu CJL, Ko YC, Lee SM and Chou P 2016 *Eye* **30** 570-6
[5] Fares U, Al-Aqaba MA, Otri AM and Dua HS 2011 *European Ophthalmic Review* **5** 50-5

[6] Zuberbuhler B, Tuft S, Gartry D and Spokes D 2013 *Corneal Surgery* (Berlin Heidelberg: Springer) p 91-134
[7] Buratto L, Slade SG and Tavolato M 2012 *LASIK: The Evolution of Refractive Surgery* (Thorofare: Slack Inc) p 1-190
[8] Krueger RR, Rabinowitz YS and Binder PS 2010 *Journal of Refractive Surgery* **26** 749-60
[9] Solomon KD, de Castro LEF, Sandoval HP, *et al.* 2009 *Ophthalmology* **116** 691-701
[10] Sutton G, Lawless M and Hodge C 2014 *Clin Exp Optom.* **97** 18-29
[11] Canovas C, van der Mooren M, Rosén R, Piers PA, Wang L, Koch DD, et al. 2015 *Journal of Cataract & Refractive Surgery* **41** 1030-1037
[12] Meyer, B., et al. 2015 *Clinical Ophthalmology* **9** 379-392
[13] Jester JV, Venet T, Lee J, Schanzlin DJ and Smith RE 1981 *Journal of Ophthalmology* **92** 172-7
[14] Salz J, Lee JS, Jester JV, Steel D, Villasenor RA, *et al.* 1981 *Ophthalmology* **88** 742-6
[15] Block MG and Block MB 1984 *Optometry and Vision Science* **61** 721-8
[16] Bao FJ, Yu AY, Kassem W, Wang QM and Elsheikh A 2011 *Journal of Refractive Surgery* **27** 345-55
[17] Kohlhaas M, Lerche RC, Draeger J, Klemm M, Ehlers N, Hjordtal J, *et al.* 1995 *European journal of Implant and Refractive Surgery* **7** 84-8
[18] Bühren J, Pesudovs K, Martin T, Strenger A, Yoon G and Kohnen T 2009 *Journal of Cataract & Refractive Surgery* **35** 846-55
[19] Recep ÖF, Çağıl N and Hasıripi H 2000 *Journal of Cataract & Refractive Surgery* **26** 1480-3
[20] Chen MC, Lee N, Bourla N and Hamilton DR 2008 *Journal of Cataract & Refractive Surgery* **34** 1886-91
[21] Hamilton DR, Johnson RD, Lee N and Bourla N 2008 *Journal of Cataract & Refractive Surgery* **34** 2049-56
[22] Chang DH and Stulting RD 2005 *Ophthalmology* **112** 1009-16
[23] Wangsupadilok B and Horatanaruang O 2011 *Journal of the Medical Association of Thailand* **94** 574.
[24] Yi W-M and Joo C-K 1999 *Journal of Cataract & Refractive Surgery* **25** 1087-92
[25] Huang D, Stulting RD, Thompson KP, Carr JD and Waring GO 1999 *Journal of Refractive Surgery* **15** 538-49
[26] Durairaj VD, Balentine J, Kouyoumdjian G, Tooze JA, *et al.* 2000 *Ophthalmology* **107** 2140-3
[27] Ogasawara K and Onodera T 2016 *Clinical Ophthalmology* (Auckland, NZ) **10** 1977-81
[28] Choudhri SA, Feigenbaum SK and Pepose JS 2005 *Journal of refractive surgery* **21** 253-9
[29] Liyanage SE and Allan BD 2012 *Journal of Cataract & Refractive Surgery* **38** 1232-9
[30] Allan BD, Hassan H and Ieong A 2015 *Journal of Cataract & Refractive Surgery* **41** 1009-17
[31] Dutt S, Steinert RF, Raizman MB and Puliafito CA 1994 *Archives of Ophthalmology* **112** 1427-36.
[32] Lee Y-C, Hu F-R and Wang IJ 2003 *Journal of Cataract & Refractive Surgery* **29** 769-77
[33] Helgesen A, Hjortdal J and Ehlers N 2004 *Acta Ophthalmologica* **82** 454-60
[34] Srivannaboon S, Reinstein DZ, Sutton HFS and Holland SP 1999 *Journal of Cataract & Refractive Surgery* **25** 1596-9
[35] Lackerbauer C-A, Grüterich M, Ulbig M, Kampik A and Kojetinsky C 2009 *Journal of Cataract & Refractive Surgery* **35** 1343-7
[36] Alio JL, Soria FA, Abbouda A and Peña-García P 2016 *British Journal of Ophthalmology* 100 626-32
[37] Chang AW, et al. 2003 *Journal of Cataract & Refractive Surgery* **29** 1204-10
[38] Lehmann EL and Casella G 1998 *Theory of Point Estimation* (2nd ed.) (New York: Springer)
[39] Hyndman RJ and Koehler AB 2006 *International Journal of Forecasting* **22** 679–88

IOP Conf. Series: Journal of Physics: Conf. Series **976** (2018) 012010 doi:10.1088/1742-6596/976/1/012010

Target Tracking Algorithm of Automotive Radar Based on Iterated Square-root CKF

Cai-ling Wang [1], Xing Xiong [1] and Hua-jun Liu [2]

[1]College of Automation, Nanjing University of Posts and Telecommunications, Nanjing 210023, China;
[2]School of Computer Science and Engineering, Nanjing University of Science and Technology, Nanjing 210094, China

Corresponding auther: jhaxing@163.com

Abstract. An iterated square-root cubature Kalman filter(ISRCKF) algorithm for target tracking of automotive radar is proposed in this paper, which inherits the fast and robust advantages of SRCKF, combines with Gauss-Newton iterative theory and design a algorithm to iterate update the measurement process. The filtering accuracy of target tracking algorithm of automotive radar can be further improved as the newest measurement is fully utilized. Monte-Carlo simulation experiments were carried out aim at the automotive radar target tracking problem , in which used the algorithm to compare with classical algorithms such as square-root unscented Kalman filter (SRUKF) and SRCKF. The experimental results shows that the overall filtering accuracy of this algorithm is much improved compare with other classical filtering algorithms, and the filtering accuracy can be improved with the increase of iteration number.

1. Introduction

Advanced driver assistance system (ADAS) is getting more and more attention in recent years[1], which is based on radar sensor technology. Accurate environmental perception is a key issue in the ADAS system, its core is to accurately detect, track and identify barriers.. Measurements obtained by automotive radar need to be non-linear transformated in ADAS, therefore, the filter in target tracking should be non-linear.

The most classical method to solve the non-linear filtering problem is the extend Kalman filter (EKF) [2-3], the EKF uses a first-order Taylor series expansion of the nonlinear function and ignores higher-order terms, then is processed by standard Kalman filtering. The EKF can lead to large errors and even become unstable when the system is highly non-linear. Unscented Kalman Filtering (UKF) is also a popular non-linear filter [4-5], the UKF obtain the statistical characteristics of random variables by sampling points with different weights (Sigma points) through non-linear functions and performing unscented transformation (UT), the filtering accuracy can reach the second order with the same performance in the EKF, and the computational efficiency is higher since it don't need to calculate the jacobian matrix. However, the filtering performance of the UKF may be poor or even divergent when the state dimension is high. Arasaratnam and Haykin proposed the cubature Kalman Filtering (CKF) [6], the CKF uses the cubature rules for those multi-dimensional integrals to approximate the recursive Bayesian estimation integrals under the Gaussian assumption. Compared with UKF, CKF is simpler to implement, has higher computational efficiency, higher filtering accuracy and numerical stability [7].

The square-root cubature Kalman filter (SRCKF) algorithm [8]is based on the CKF framework, propagates the square-root factors of the predictive and posterior error covariance to avoid matrix square-rooting operations and to preserve the positive semi-definiteness of the error covariance, which effectively improves the accuracy and stability of the filtering.

In the target tracking of automotive radar, the process is affected by all kinds of noise and non-linear transformation, which requires fastness, robustness and accuracy of the filter. Based on the above research, an iterative square-root cubature Kalman filter (ISRCKF) algorithm for target tracking of automotive radar is proposed, which draws the advantages of SRCKF algorithm and iteratively optimizes the process of SRCKF measurement and updating by Gauss-Newton iterative method.

2. Description of system model

In ADAS system, the state of target can be defined as a 6-dimensional vector in the $x-y$ Cartesian coordinate system:$X_k = [x, \dot{x}, \ddot{x}, y, \dot{y}, \ddot{y}]^T$, the components of the state vector respectively indicate the target's distance, velocity and acceleration on the horizontal and vertical axes at time instant k. The measurement by a single period of radar is $z_k = (r, a, v)^T$, meaning distance r, azimuth a and radial velocity v of the target, respectively. The kinematics model and measurement model in target tracking of automotive radar are shown in equation (1):

$$\begin{cases} X_k = F \cdot X_{k-1} + w_{k-1} \\ z_k = h(X_k) + v_k \end{cases} \tag{1}$$

Where k is the discrete time; X_k is the state vector of the system at time instant k; z_k is the measurement; F and $h(\cdot)$ are the state equation and measurement function of the system; w_{k-1} and v_k denote the process and measurement Gaussian white noise, which with zero means and covariance Q_{k-1} and R_k, respectively, $\{w_{k-1}\}$and $\{v_k\}$ are mutually uncorrelated. The motion of target can be assumed to be a normally accelerating motion. Therefore, the state equation F of the system model is based on the normal acceleration (CA) model [9-10], and the time interval dt between two measurements of the radar as the equation of state parameters, the F can be expressed as:

$$F = \begin{bmatrix} 1 & dt & \frac{1}{2}dt^2 & 0 & 0 & 0 \\ 0 & 1 & dt & 0 & 0 & 0 \\ 0 & 0 & 1 & 0 & 0 & 0 \\ 0 & 0 & 0 & 1 & dt & \frac{1}{2}dt^2 \\ 0 & 0 & 0 & 0 & 1 & dt \\ 0 & 0 & 0 & 0 & 0 & 1 \end{bmatrix} \tag{2}$$

The process noise covariance matrix Q is initialized to $Q = \mathrm{diag}[\sigma_x \; \sigma_{\dot{x}} \; \sigma_{\ddot{x}} \; \sigma_y \; \sigma_{\dot{y}} \; \sigma_{\ddot{y}}]$

The measurement function $h(\cdot)$ completes the transformation of the state vector X_k to the measurement vector z_k, as defined below:

$$h(X_k) = \begin{bmatrix} \sqrt{{x_k}^2 + {y_k}^2} \\ \arctan\,(y_k/x_k) \\ (x_k\dot{x}_k + y_k\dot{y}_k)/\sqrt{{x_k}^2 + {y_k}^2} \end{bmatrix} \tag{3}$$

The measurement noise covariance matrix R is initialized to $R = \mathrm{diag}[\sigma_r \; \sigma_a \; \sigma_v]$.

3. Theory of iterative square-root CKF

In a nonlinear discrete dynamic systems described in equation(1), $\hat{x}_{\bar{k}} \sim N(x_k, \hat{P}_{\bar{k}})$, $z_k \sim N(h(x_k), R_k)$; From the Bayesian perspective we have $P(x_k|z_k) = \eta P(z_k|x_k)P(x_k)$, where $\eta = P_{z_k}^{-1}$ amounts to a normalizing factor, $P(x_k)$ is the prior, and $P(z_k|x_k)$ is the likelihood:

$$P(x_k) = \frac{1}{\sqrt{(2\pi)^n|\hat{P}_{\bar{k}}|}} \exp\left[-\frac{1}{2}(x_k - \hat{x}_{\bar{k}})^{\mathrm{T}}\hat{P}_{\bar{k}}^{-1}(x_k - \hat{x}_{\bar{k}})\right] \tag{4}$$

$$P(z_k|x_k) = \frac{1}{\sqrt{(2\pi)^l|R_k|}} \exp\left[-\frac{1}{2}(z_k - h(x_k))^{\mathrm{T}}R_{\bar{k}}^{-1}(z_k - h(x_k))\right] \tag{5}$$

Defining the likelihood surface function (LSF) as

$$J(x_k) = \frac{1}{2}\left[(x_k - \hat{x}_{\bar{k}})^{\mathrm{T}}\hat{P}_{\bar{k}}^{-1}(x_k - \hat{x}_{\bar{k}}) + (z_k - h(x_k))^{\mathrm{T}}R_{\bar{k}}^{-1}(z_k - h(x_k))\right] \tag{6}$$

Obtaining the maximum likelihood estimate is equivalent to solving the minimum value of $J(x_k)$. Gauss-Newton nonlinear iterative method is introduced to linearize the measurement equation to obtain the updated measurement iteration equation of $J(x_k)$[11]:

$$\hat{x}_k^{(i+1)} = \hat{x}_{\bar{k}} + \hat{P}_{\bar{k}}(H_k{}^{(i)})^T(H_k{}^{(i)}\hat{P}_{\bar{k}}(H_k{}^{(i)})^T + R_k)^{-1}[z_k - h(\hat{x}_k^{(i)}) - H_k{}^{(i)}(\hat{x}_{\bar{k}} - \hat{x}_k^{(i)})] \tag{7}$$

Where $\hat{x}_k^{(i)}$ is the state estimation at the i-th iteration, $H_k{}^{(i)}$is the jacobian matrix of $h(\hat{x}_k^{(i)})$, we consider the iterative filter gain $K_k^{(i)} = \hat{P}_{\bar{k}}(H_k{}^{(i)})^{\mathrm{T}}(H_k{}^{(i)}\hat{P}_{\bar{k}}(H_k{}^{(i)})^{\mathrm{T}} + R_k)^{-1}$; the terms $(H_k{}^{(i)}\hat{P}_k(H_k{}^{(i)})^{\mathrm{T}} + R_k)$ and $\hat{P}_k(H_k{}^{(i)})^{\mathrm{T}}$are innovation covariance and cross covariance, they are expressed as $P_{zz}^{(i)} = H_k{}^{(i)}\hat{P}_{\bar{k}}(H_k{}^{(i)})^{\mathrm{T}} + R_k$,$P_{xz}^{(i)} = \hat{P}_{\bar{k}}(H_k{}^{(i)})^{\mathrm{T}}$, $P_{zz}^{(i)}$ and $P_{xz}^{(i)}$ are obtained by linearizing the measurement function $h(\cdot)$ with a first-order taylor expansion the linearized error is introduced due to high-order terms being truncated. To decrease the propagated linearized error, we recalculate innovation covariance and cross covariance based on the SRCKF in the following way:

$$Z_{j,k}^{(i)} = h(X_{j,k}^{(i)}) \tag{8}$$

$$\hat{z}_k^{(i)} = \sum_{j=1}^{m} w_j Z_{j,k}^{(i)} \tag{9}$$

$$\chi_k^{(i)} = [X_{1,k}^{(i)} - \hat{x}_k^{(i)}, X_{2,k}^{(i)} - \hat{x}_k^{(i)}, \dots, X_{m,k}^{(i)} - \hat{x}_k^{(i)}]/\sqrt{m} \tag{10}$$

$$\eta_k^{(i)} = [Z_{1,k}^{(i)} - \hat{z}_k^{(i)}, Z_{2,k}^{(i)} - \hat{z}_k^{(i)}, \dots, Z_{m,k}^{(i)} - \hat{z}_k^{(i)}]/\sqrt{m} \tag{11}$$

$$S_{zz,k}^{(i)} = qr\{[\eta_k^{(i)}, \sqrt{R_k}]\} \tag{12}$$

$$P_{xz,k}^{(i)} = \chi_k^{(i)}\eta_k^{(i)} \tag{13}$$

Where $X_{j,k}^{(i)}$ in equation (8) is the cubature points, $m = 2n$, $w_j = 1/m$ is the weight of the cubature points[6]; Derived from the above, the state estimation in equation(7) can be derived as follows :

$$\hat{x}_k^{(i+1)} = \hat{x}_{\bar{k}} + P_{xz,k}^{(i)}((S_{zz,k}^{(i)})^{\mathrm{T}}S_{zz,k}^{(i)})^{-1}[z_k - h(\hat{x}_k^{(i)}) - (P_{xz,k}^{(i)})^{\mathrm{T}}\hat{P}_{\bar{k}}^{-1})(\hat{x}_{\bar{k}} - \hat{x}_k^{(i)})] \tag{14}$$

4. The algorithm of iterative square-root CKF

Combined with the Gauss-Newton iterative method, the improved innovation covariance and cross covariance algorithm, the specific process of ISRCKF algorithm is as follows:

Step 1: Using the three-order cubature rule to evaluate the cubature points and weights:

$$\xi_j = \sqrt{(\mathrm{m}/2)}[1]_j, w_j = 1/m, j = 1, \dots, m \tag{15}$$

Where n is the dimension of state vector, we denote the n-dimensional unit vector as $e = [1,0,\dots,0]^{\mathrm{T}}$, use the $[1]$ to represent a complete fully symmetric set of points that can be obtained by permutating and changing the sign of the generator e in all possible ways. The $[1]_j$ is denoted the j-th point from the set $[1]$.

Step 2: Time update;

IOP Conf. Series: Journal of Physics: Conf. Series **976** (2018) 012010 doi:10.1088/1742-6596/976/1/012010

Assuming the posterior density function $p(x_{k-1}|z_{k-1}) = N(x_{k-1}; \hat{x}_{k-1}, P_{k-1})$ at $k-1$ instant time is known, where $P_{k-1} = (S_{k-1})^{\mathrm{T}}(S_{k-1})$

1) Calculate the cubature points:

$$X_{j,k-1} = S_{k-1}\xi_j + \hat{x}_{k-1} \tag{16}$$

2) Propagate the cubature points through the state equation:

$$X_{j,k}^{*} = f(X_{j,k-1}) \tag{17}$$

3) Evaluate the predicted state and square-root of the predicted covariance

$$\hat{x}_{\bar{k}} = \sum_{j=1}^{m} w_j X_{j,k}^{*} \tag{18}$$

$$\chi_k^{*} = [X_{1,k}^{*} - \hat{x}_{\bar{k}}, X_{2,k}^{*} - \hat{x}_{\bar{k}}, \dots, X_{m,k}^{*} - \hat{x}_{\bar{k}}]/\sqrt{m} \tag{19}$$

$$\hat{S}_{\bar{k}} = \mathrm{qr}\{[\chi_k^{*}, \sqrt{Q_k}]\} \tag{20}$$

Step 3: Measurement update;

Measurement update is an iterative process with $\hat{x}_{\bar{k}}$ and $\hat{S}_{\bar{k}}$ as the initial value, the initial covariance matrix $\hat{P}_{\bar{k}} = (\hat{S}_{\bar{k}})^{\mathrm{T}}\hat{S}_{\bar{k}}$. The estimation of state and square-root of covariance obtained by i-th iteration are $\hat{x}_k^{(i)}$ and $\hat{S}_k^{(i)}$, $\hat{x}_k^{(0)} = \hat{x}_{\bar{k}}$, $\hat{S}_k^{(0)} = \hat{S}_{\bar{k}}$, for $\mathrm{i} = 1,2,\dots,\mathrm{N_{iter}}$:

1) Calculate the updated cubature points

$$X_{j,k}^{(i-1)} = \hat{S}_k^{(i-1)}\xi_j + \hat{x}_k^{(i-1)} \tag{21}$$

2) Use equations (8-13) to evaluate the square-root of innovation covariance $S_{zz,k}^{(i-1)}$ and cross-covariance $P_{xz,k}^{(i-1)}$
3) Evaluate the iterative filter gain $K_k^{(i-1)} = P_{xz,k}^{(i-1)}((S_{zz,k}^{(i-1)})^{\mathrm{T}} S_{zz,k}^{(i-1)})^{-1}$
4) Evaluate the state estimation and the square-root of the covariance

$$Z_k^{(i-1)} = h(\hat{x}_k^{(i-1)}) \tag{22}$$

$$\hat{x}_k^{(i)} = \hat{x}_{\bar{k}} + K_k^{(i-1)}[z_k - Z_k^{(i-1)} - (P_{xz,k}^{(i-1)})^{\mathrm{T}}\hat{P}_{\bar{k}}^{\;-1}(\hat{x}_{\bar{k}} - \hat{x}_k^{(i-1)})] \tag{23}$$

$$\hat{S}_k^{(i)} = qr([\chi_k^{(i-1)} - K_k^{(i-1)}\eta_k^{(i-1)}, K_k^{(i-1)}\sqrt{R_k}]) \tag{24}$$

5) Termination condition of iteration: $\left\|\hat{x}_k^{(i)} - \hat{x}_k^{(i-1)}\right\| \le \varepsilon$ or $i = N_{iter}$

Where ε and $\mathrm{N_{iter}}$ is the predetermined threshold and the maximum number of iterations. The iteration number is assumed to be N when the iteration is terminated; The ultimate estimation of state and square-root of covariance at the k time instant are $\hat{x}_k = \hat{x}_k^{(N)}$, $\hat{S}_k = \hat{S}_k^{(N)}$.

5. Simulation

Totally two experiments of 50 Monte-Carlo simulations were carried out in the simulation experiment. The first experiment compares the tracking filtering accuracy of SRUKF [12], SRCKF and ISRCKF. The second compares the iteration number N_{iter} of ISRCKF to the filtering accuracy influences. The same Gaussian white noise is added into each experiment, the mean square-root error of the distances, azimuths and velocities are 2.0*m*, 1.0° and 0.2 *m/s*. The data update rate of radar is 20 Hz and the target is tracked for 30*s*. The initial state of the target is: $[10, 0.5, -0.1, 10, 2, -0.01]^{\mathrm{T}}$.

In experiment 1, the iteration number N_{iter} of ISRCKF is set to 2. Figure 1 and 2 respectively shows the MSRE curves of radial distance and azimuth of 3 filtering algorithms obtained from 50 Monte-Carlo simulations. The filtering accuracy of each algorithm is quantitatively analyzed, table 1 shows the AMSRE of $[x, \dot{x}, \ddot{x}, y, \dot{y}, \ddot{y}]$ and$[r, a, v]$. From table 1 we can see that the error of ISRCKF in every component are much lower than that of SRUKF and SRCKF.

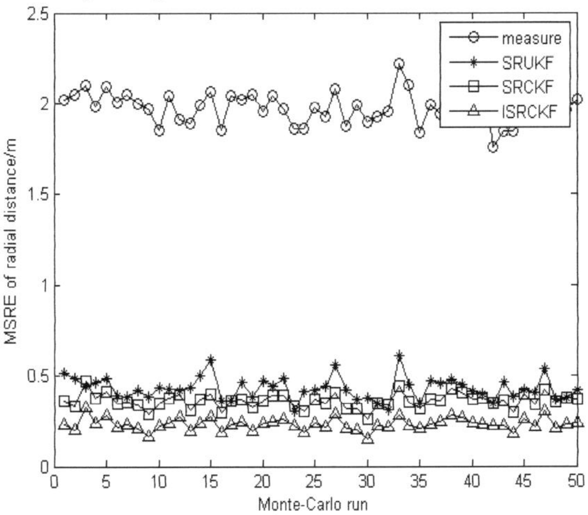

Figure 1. MSRE curves of radial distance for various filters.

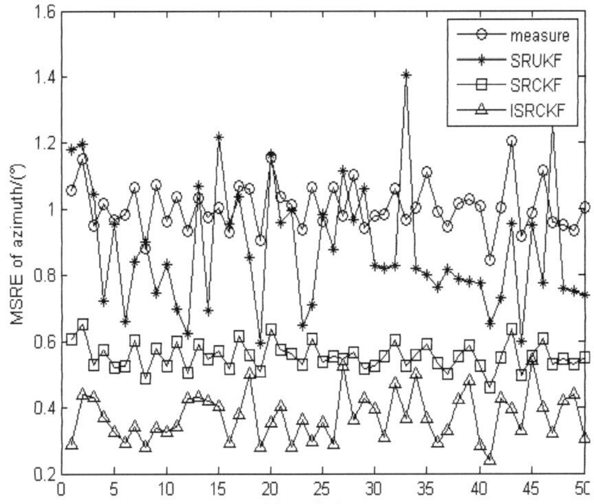

Figure 2. MSRE curves of azimuth for various filters.

Table 1. Accumulated mean square-root error (AMSRE) of various algorithms.

AMSRE	$[x, \dot{x}, \ddot{x}]$	$[y, \dot{y}, \ddot{y}]$	$[r, a, v]$
measurement	[0.746, 0.148,~]	[1.815, 0.292,~]	[2.005, 0.998, 0.200]
SRUKF	[0.308, 0.140, 0.004]	[0.404, 0.107, 0.004]	[0.427, 0.832, 0.022]
SRCKF	[0.280, 0.113, 0.003]	[0.336, 0.049, 0.004]	[0.363, 0.548, 0.017]
ISRCKF	[0.116, 0.074, 0.002]	[0.212, 0.042, 0.003]	[0.233, 0.380, 0.013]

In experiment 2, the iteration number N_{iter} of ISRCKF is set to 1, 2, 3 and 4, respectively. Figure 3 and 4 respectively shows the MSRE curves of radial distance and azimuth with different iteration numbers obtained from 50 Monte-Carlo simulations. The filtering accuracy of each algorithm is quantitatively analyzed, table 2 shows the AMSRE of $[x, \dot{x}, \ddot{x}, y, \dot{y}, \ddot{y}]$ and $[r, a, v]$. From table 2 we can see that the filtering accuracy of ISRCKF improved with the increase of iteration number, that owing to the newest measurement is fully utilized during the iteration.

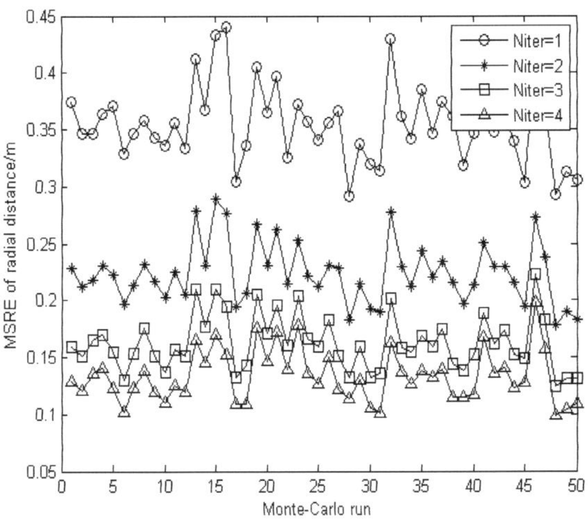

Figure 3. MSRE curves of radial distance for ISRCKF with various iteration numbers.

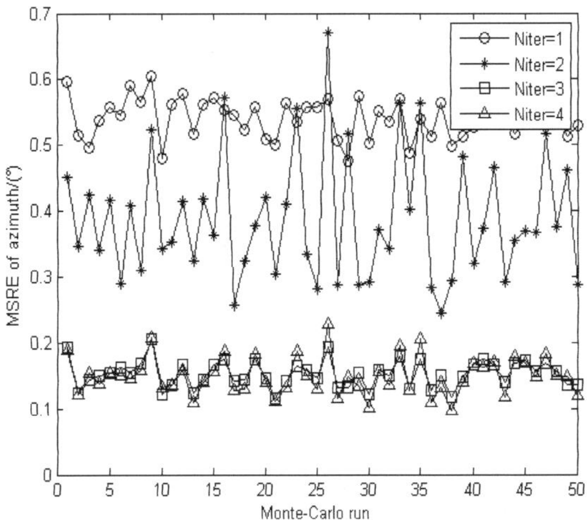

Figure 4. MSRE curves of azimuth for ISRCKF with various iteration numbers

IOP Conf. Series: Journal of Physics: Conf. Series **976** (2018) 012010 doi:10.1088/1742-6596/976/1/012010

Table 2. Accumulated mean square-root error (AMSRE) of ISRCKF with various iteration numbers

AMSRE	$[x, \dot{x}, \ddot{x}]$	$[y, \dot{y}, \ddot{y}]$	$[r, a, v]$
N_{iter}=1	[0.281, 0.113, 0.003]	[0.336, 0.048, 0.003]	[0.362, 0.549, 0.017]
N_{iter}=2	[0.116, 0.074, 0.002]	[0.210, 0.041, 0.004]	[0.232, 0.374, 0.013]
N_{iter}=3	[0.095, 0.062, 0.001]	[0.150, 0.038, 0.003]	[0.170, 0.156, 0.011]
N_{iter}=4	[0.077, 0.060, 0.001]	[0.123, 0.037, 0.003]	[0.139, 0.149, 0.010]

6. conclusion

Based on SRCKF and Gauss-Newton iteration theory, an iterative square-root cubature Kalman filter (ISRCKF) algorithm is proposed for target tracking of automotive radar. The ISRCKF use the same method as SRCKF to predict the state and the square-root of covariance. In the measurement update, the Gauss-Newton nonlinear iteration method is used to perform iterative updating until the difference between the state estimation obtained by two consecutive iterations is less than the predetermined threshold or the iteration number reaches the predetermined maximum number, the latest measurement is fully utilized during the iterations to achieve a good filtering effect. Under the simulation of the target tracking of automotive radar, the experimental results show that the filtering accuracy of ISRCKF algorithm is much better than that of the classical SRUKF and SRCKF algorithms, and the filtering accuracy increases with the increase of iteration number within a certain range。

References

[1] Mosquet X, Andersen M, Arora A. A roadmap to safer driving through advanced driver assistance systems[J]. *Auto Tech Review*, 2016, 5(7): 20-25.

[2] Bolognani S, Tubiana L, Zigliotto M. Extended Kalman filter tuning in sensorless PMSM drives[J]. *IEEE Transactions on Industry Applications*, 2003, 39(6): 1741-1747.

[3] Gustafsson F, Hendeby G. Some relations between extended and unscented Kalman filters[J]. *IEEE Transactions on Signal Processing*, 2012, 60(2) : 545-555.

[4] Julier S J, Uhlman J K.Unscented Filtering and Nonliner Estimation[J]. *Proceedings of the IEEE*, 2006, 13(2) : 99-108.

[5] Julier S J, Uhlman J K, Durrant-Whyte H F. A New Method for the Nonlinear Transformation of Means and Covariances in Filters and Estimators[J]. *IEEE Transactions on Automatic Control*, 2000, 45 (3) : 477-482..

[6] Arasaratnam I, Haykin S. Cubature Kalman Filters[J]. *IEEE Transactions on Automatic Control*, 2009,54 (6) :1254-1269.

[7] SUN Feng, TANG Lijun. Estimation precision comparison of cubaturekalman filter and unscented kalman filter[J]. *Control and Decision*, 2013, 28(2) : 303-308.

[8] Jing Mu, Yuanli Cai. Square Root Cubature Kalman Filter Algorithm and Application. *Ordnance Industry Automation*. 2011; 30(6): 11-14

[9] Mazor E, Averbuch A, Bar-Shalom Y, et al. Interacting multiple model methods in target tracking: a survey[J]. *IEEE Transactions on aerospace and electronic systems*, 1998, 34(1): 103-123.

[10] Han S K, Ra W S, Whang I H, et al. Linear recursive automotive target tracking filter for advanced collision warning systems[J]. *Applied Mathematics & Information Sciences*, 2014, 8(3): 1145-1151.

[11] Sibley G, Sukhatme G S, Matthies L H. The Iterated Sigma Point Kalman Filter with Applications to Long Range Stereo[C]//*Robotics: Science and Systems*. 2006, 8(1): 235-244.

[12] Zhinong W, Guoqiang S, Bo P. Application of UKF and SRUKF to power system dynamic state estimation[J]. Proceedings of the CSEE, 2011, 31(16): 74-80.

A Case Study of Air Cleaner by the Intelligent Interaction and Emotion

Huai Cao[1] and Yuwen Sun[1]

[1]Huazhong University of Science and Technology, 430074 Wuhan, P.R.China

Email: caohuai@hust.edu.cn;18652093692@163.com

Abstract. The pure and fresh air can not only contribute to our physical and mental health, but also can be beneficial to ease the pressure and relax the mood. The vertical intelligent air cleaner can remove the harmful gases from the air and absorb the suspended particles in the air, especially all kinds of the bacteria and viruses. The air cleaner is good for improving the air quality of the indoor and maintaining the health of the people. The designing of the vertical air cleaner is as follows: The designing of the vertical intelligent make full use of the developed air purification technology. The smart home is inserted into the work. Simultaneously, in the aspect of the design of intelligent products, the intelligent interactive processes are scientifically planned. Moreover, the emotional design and the user experience are fully considered, which can enhance the comprehensive design ability.

1. Introduction

With the development of industrialization, the air pollution is becoming more and more serious. There are nine major chemicals in the air which can cause serious harm to human health. Excessive absorption of these chemical poisons can lead to multiple diseases and life-threatening conditions. The air pollution sources indoor mainly come from four aspects, such as the cement, the paint, the paint and furniture, as well as the household appliances and other indoor equipment, as well as the heating and cooking process. Persons live and work in the indoor environment in the most of the time of the day. The air quality indoor is the guarantee of healthy living, so keeping the air fresh and healthy can improve and enjoy our life.

There are major of reports about developing the air cleaner products. A new bench test for quantitative laboratory of air cleaner silencing is introduced [1]. With the development of date sample and intelligent-control technology, the hardware and software design and application of the new kind of intelligent digital air cleaner is applied [2]. Using a numerical simulation method and experimental method, the optimal design of acoustic performance for air-cleaner was presented and evaluated.[3] The air purifier is designed with a single chip which can absorb, decompose or transform all kinds of air pollutions.[4]Forming parts using a single kind of recyclable material is the most effective way to satisfy the requirement which is eco-friendly to the environment and a new air cleaner filter is developed.[5] A finite element model an automotive air cleaner is established with FEA model.[6]

In this paper, the innovative design of a vertical intelligent healthy air machine is completed, and the existing patent of air purification core technology is fully adopted. At the same time, in the intelligent design of the product, the intelligent interaction flow can be scientifically planned, and the emotional design and user experience should be fully considered, and the comprehensive design capability will be improved.

2. Social needs

2.1 Use the crowd

Women, especially pregnant women. Because women have a thicker fat, benzene absorbs easily and is quickly stored in fat, so women should pay more attention to benzene. Women should avoid contact with decoration during pregnancy.

Children. Children's bodies are in developmental stages, the immune system is more fragile, and their breathing rate is 50 percent higher than that of adults. It makes them more vulnerable to indoor air pollution damage for a long time indoor. The body of the Children is in the critical period of growth and development. With a long time inhalation of smoke, harmful gases, bacteria and viruses, not only it is easy to induce a variety of diseases, but also it attack the children's body function by the chronic effect, thereby affecting the normal development of height and intelligence.

The elderly. As the body enters its old age, various indicators of the body have gradually decreased, and it is easy to be affected by air quality.

Patients with respiratory diseases. According to the survey, 68 percent of the human body's illnesses are related to environmental air pollution. The most common human disease is the respiratory infections, and its symptoms can range from an implicit infection to a life-threatening illness.

People living in cities or areas where air quality is poor.

2.2 Applicable places

a. Newly renovated or renovated residence.
b. The residence of the elderly, children, pregnant women and newborn.
c. Home to asthma, allergic rhinitis and pollen allergens.
d. The residence of keeping pets and livestock. 5.
e. Dwelling place that is more closed or affected by second-hand smoke.
f. Hotels, public places.
g. Living in high quality.
h. Hospital which can reduce infection, prevent transmission of disease.

3. Standard and core technical theory

3.1 Standard

a. Cleanliness. The clearance rate of PM2.5 and PM10 are above 95%. Effective intervention of radioactive radon is achieved by 99.9% clearance.

b. Temperature. Winter: 16-24 ℃; Summer: 22-28 ℃.

c. Freshness. The new air volume is above 30 cubic meters per hour. The amount of oxygen is comparable to that of local outdoor level. The content of 12 poisonous and harmful chemical gases will be reduced to the national standard, and the content of unpleasant odor will be reduced by more than 95 %.

d. Humidity. Winter: 30-60% ; summer: 40-80%

e. Bacteria. The bacterial content is reduced to the national standard.

3.2 The Key Theory

3.2.1 The core theory. The core theory include purify and replace, which continuously the fresh clean air is injected, and the turbidity gas is expelled.

3.2.2 The Core Technologies. The Positive Pressure Shielding: Air is a fluid of gaseous substances. The room is viewed as a unclosed container, and constantly fresh clean air is injected to the room. The air indoor will continue to increasing under the positive pressure condition upper and it will force the interior air pollution to outside leakage through doors and windows, exhaust pipe and channel. Not only it can make indoor air pollution gradually dilute and extrude outdoor side by side, but also can prevent the untapped air outdoor returning home freely, with it form the positive pressure blocking for the outdoor air. As a result, it ensure the long-term stability of the indoor air quality.

The Photoelectric Purification System: Because of air ionization characteristics, the photoelectric purification system release millions of high concentration negative ions in every second to achieve environment best, and make the dust particles in the air negatively charged and then captured and adsorbed by the high magnetic electrostatic field.

3.3 The Purification Process

The outdoor fresh air is sucked into the indoor air quality optimization machine. The new air is treated clean from the inside of the machine, and the removal rate is above 95%, which is better than the national standard. The clean air is sterilized (the rate of bacterial inactivation is above 97%), which the total number of colonies is less than the standard value of the standard. The detoxification air continues to be treated with the odor, and the contents of the 13 pollutants and the unlisted TVOC contents are lower than the national standard. The above optimization air can meet the requirements of physical parameters in the national standard after the temperature compensation and humidification treatment. The air quality is further improved by adding negative ions and natural flavors and make it a healthy air. Then the healthy air will be continuously injected into the indoor to meet the new air consumption requirements of more than 30M3 per person per hour. The healthy air gradually dilute and squeeze out the indoor air, thus achieving a comprehensive optimization and management of indoor air quality.

4. The Project Design

4.1 The Product Design

4.1.1 The Design Concept. The upper part of the display board is a common mess in some of the existing rooms, containing of PM2.5, pollen, allergens, dust, fungi, smoke, formaldehyde and benzene. All these things are harm of our bodies all the time. The mess is filtered through the filter core and then filtrated through the linen or non-woven fabrics. The linen or non-woven fabrics are easy to absorb the water and easy to dry, adding incense and humidification directly to the cloth cover. The fresh air spread into the room through the cloth cover with the fragrance and water. The whole machine works by using a fan at the bottom to suck the dirty air in, through the triple purification and finally filtered through the linen in figure 1.

4.1.2. The Appearance Design. The appearance make persons be close to nature indoors, and feel new and fresh. The product main material which are as far as possible environmental and recycled are selected for linen and stainless steel. These can serve as the adornment and furniture.

4.1.3. The Structure Design. The whole is composed of the universal wheel, the fan, the bracket, the filter core, the lamp and the cloth cover in figure 2.

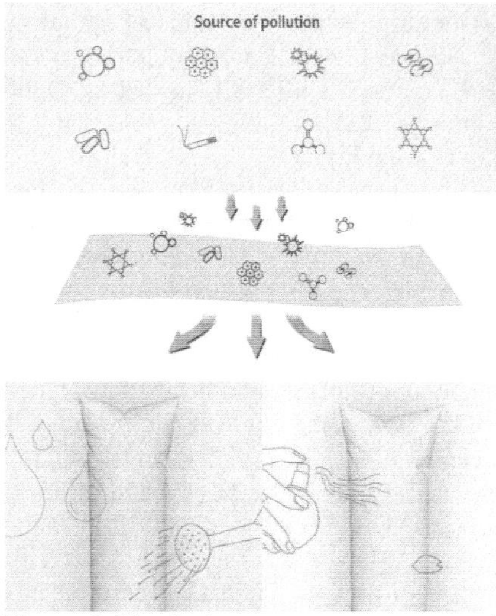

Figure 1. The design concept

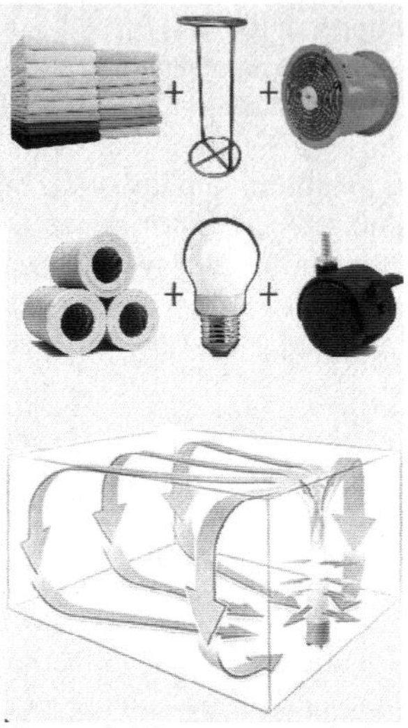

Figure 2. The structure design

4.2 The Interface Design

A APP is designed, which monitors fully intelligently the quality of the air indoor, and the data is collected automatically, and analyzes by the powerful software ,timely get the analysis conclusions and command execution device to make appropriate decisions. It can take different automatic control plans and implementations, according to the air pollution type and pollution levels indoor, respectively.

a. Equipped with two control methods: the wireless remote control and the keyboard control. Users can choose to use them at will.

b. A variety of operating modes for the user to select such as the intelligent control and manual setting control meet the needs of children, the older who are easy to operate, and take care of special user's personalized need.

c. The display function can show the air temperature, humidity, freshness and cleanliness, which the indoor air quality status are well known.

e. With functions of the self-diagnosis and prompt Settings, the human-machine interactive dialog easily implement on the control interface. When there is abnormal product running state, the computer will automatically show the problem, and sing to remind users to timely treatment.

Figure 3. The interface of APP

In the main interface in Figure 3, there are local weather conditions. If there are several air purifiers in the home, they can be displayed in the same account. The air qualities of each region are tested and recommendations are given at the same time. The interface of the each region consist of the current state, the history record, the time switch. The current state is to tell the user whether to need to open air purification machine and the history can be found such as the daily basis, statistical results, and then the diagrams are draw. According to the analysis result, the timer switch will start. It can be activated before getting home and when the child the old are at home. The wind direction can also be adjusted after opening, and the phone can be controlled remotely.

5. The Application Effect

It is proved that the air purifier can provide healthy air for a family (about 80 to 160 square meters), and the power consumption is only about 50W in the normal operation. In the case of maintaining long - term cleaning, the power consumption is only about 30W.The daily cost of electricity is about 25RMB per month. Each family usually only chooses to install one, and any room can be easily converted.

This air purifier is eco-friendly. The filter element and the cloth cover can be replaced at will and it is very convenient of cleaning the cloth cover regularly. At the same time, the APP is designed to connect with the mobile phone to realize remote control and facilitate people's life.

6. The Conclusion and Extension

The vertical intelligent healthy air machine can effectively remove the harmful gas and adsorb the suspended particles in the air, especially the various bacteria and viruses. The vertical healthy air purifier can produce the fresh air. Pure and fresh air is not only good for your physical and mental health, but it can also help relieve stress and relax.

The vertical intelligent health air machine, which fully adopts the existing patent of air purification core technology, integrates smart home modules into the works. At the same time, in the intelligent design of the product, the scientific planning intelligent interaction process, and the emotional design and the user experience are fully considered.

Reference

[1] Sherburn, Peter E. Air Cleaner Design-Present and Future. February 1,1969 by SAE International in US.
[2] Li Jingguang, Luo Fei, Pan Zhifeng. The design of a new kind of intelligent air cleaner. Embedded System and SOC, n3,2007.
[3] Chi Liu, Zhi-yong Hao, Xin-rui Chen. Optimal design of acoustic performance for automotive air-cleaner. Applied Acoustics, v 71, n 5, May 2010.
[4] Li Mei Wen. Design of Automatic Air Purification Device. Applied Mechanics and Material, v644-650,2014.
[5] Oda Kouichi, Honda Minoru. Development of a new air cleaner filter. SAE Technical Papers,2001.
[6] Fang Zhang, Hong Zhou. The model analysis and optimum structure design of automobile air cleaner. Advanced Materials Research, v 516-517,2012

IOP Conf. Series: Journal of Physics: Conf. Series **976** (2018) 012012 doi:10.1088/1742-6596/976/1/012012

Electromagnetic Compatibility Design of the Computer Circuits

Hong Zitai

School of Instrument Science and Engineering, Southeast University, Nanjing, 211189, China

Email:407103478@qq.com

Abstract. Computers and the Internet have gradually penetrated into every aspect of people's daily work. But with the improvement of electronic equipment as well as electrical system, the electromagnetic environment becomes much more complex. Electromagnetic interference has become an important factor to hinder the normal operation of electronic equipment. In order to analyse the computer circuit compatible with the electromagnetic compatibility, this paper starts from the computer electromagnetic and the conception of electromagnetic compatibility. And then, through the analysis of the main circuit and system of computer electromagnetic compatibility problems, we can design the computer circuits in term of electromagnetic compatibility. Finally, the basic contents and methods of EMC test are expounded in order to ensure the electromagnetic compatibility of equipment.

1. Introduction

Computer is a digital device which processes data and information. Generously, information is transmitted and processed in the form of an electromagnetic signal. With the development of information technology and the popularization of microcomputers, computer has become an electronic device which is widely used in information processing and information transmission. Due to the special structure of the computer and the way to work, it transmits electromagnetic wave inevitably. The electromagnetic radiation will jeopardize the surrounding in three ways. The first one is that it is detrimental to the health of those people who are around the computer; the second one is that it will cause the formation of electromagnetic interference; the third is the information leakage. Therefore, the research on the compatibility of computer and electromagnetic environment has become a global concern.

This article mainly includes the significance of researching the computer electromagnetic compatibility, the concept of the computer electromagnetic compatibility, EMC design of the computer and the electromagnetic compatibility test. The structure of this paper is shown in Figure 1. This paper analyzes and put forward some measures about electromagnetic compatibility design of the computer circuits from these aspects.

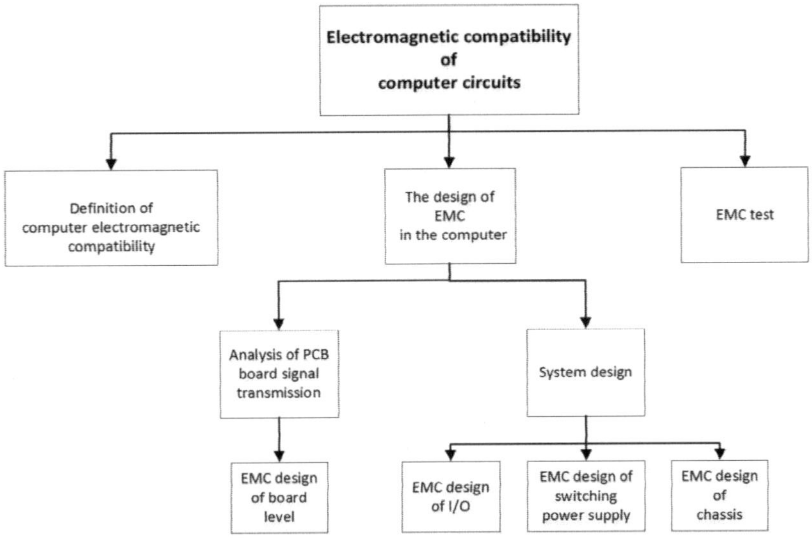

Figure 1. The structure of this paper

2. Conception of computer electromagnetic compatibility

Electromagnetic compatibility (EMC) is the branch of science and engineering concerned with the design and operation of equipment in a manner that makes them immune to certain amounts of electromagnetic interference, while at the same time keeping equipment generated interference within certain specified limits. EMC must be regarded as an issue that affects all aspects of design: electrical, electronic, and mechanical. It cannot be adequately addressed in isolation. Typically, a complete design may consist of a number of subsystems that interact with each other through signal and power cables and through reactive (capacitive and/or inductive) or radiative mechanisms [1].

The hardware of the computer usually contains digital circuits, which is predominant, and analog circuit. The crosstalk among outside electromagnetic radiation, internal devices and the various transmission channel generates interference or threaten the stability and reliability of computer.

3. EMC design of the computer

This section focuses on the analysis of the electromagnetic radiation and electromagnetic compatibility in the main part of computer system as well as the computer board level, and put forward the pertinence measures in computer design to further increase the EMC of the overall system.

3.1. EMC design of computer board level

The mainboard is a large circuit board with sockets and memory slots, and a large scale integrated circuit chip called a chipset. The mainboard is one of the main components of computer generating electromagnetic leakage. Then, the computer mainboard is a complex PCB board. PCB board refers to the printed circuit board, which is the electronic components of the electrical connections provider. The analysis of the factors affecting the PCB board and EMC design of the PCB board will be discussed below.

3.1.1. Factors affecting high speed PCB signal transmission

1) Interconnect delay is the time at which the signal passes from the output end of the drive device to the receiving end. The interconnect delay depends on the size of the signal transmission rate, interconnect length, and the capacitive load that causes the rise time to elongate.

2) When the signal energy passes through the conductor and is not absorbed at the terminal of the conductor, some of the energy will be reflected back to the conductor, resulting in reflected noise. In

high speed systems, reflected noise causes a change in the state of the logic, which often results in triggering of errors.

3) Crosstalk between signal lines is a major concern in high speed PCB designs since it can cause electromagnetic interference issues apart from conventional signal integrity (SI) problems. Since crosstalk between signal lines increases with frequency, it becomes increasingly important to consider the impact of crosstalk on EMC for very high speed PCBs. As a rule of thumb, when 100μV appears on an unshielded wire leaving a shielded enclosure, the fields at 10m will violate the EMC standards for computer hardware. Thus, for meeting EMC standards and avoiding re-design, it is imperative to factor in the impact of crosstalk on EMC even at the early stages of PCB design [2].

3.1.2. EMC design of PCB board for high speed digital system

1) Select and use multilayer PCB reasonably

a. Do not route clock and high speed signal on the surface, in case if the requirement is not meet, restrict routing length to minimum 2mm - 5 mm.

b. For all long (length > 2mm-5mm) surface signals ensure that these signals do not run parallel to any clock/high speed traces in the inner layers of board.

c. Avoid routing of power plane associated with clock circuitry.

d. Avoid routing of long floating (un-terminated)/high impedance terminated traces.

e. Do not cut/punch GND plane below the outer layer routing [3].

f. Separating digital circuits from analog circuits, because high-speed digital signals have a wide spectrum, which are the primary source of interference. Thus it must be arranged separately from the sensitive circuitry.

g. The wiring layer should be arranged adjacent to the whole metal plane. This arrangement is designed to produce energy offsets.

2) The arrangement of components

Components should be grouped on the board. The purpose is to partition the space on the printed board, so as to ensure that the components of each group do not interfere with each other in space.

3) The layout of the ground wire

The first consideration of the ground line is to set ground wire in digital circuits and analog circuits relatively according to the different power supply voltage.

4) The arrangement of the power line and filtering.

There will be a large number of transient supply electricity in the PCB circuit of the high-speed digital system, which will cause EMI. Using decoupling capacitor and reducing the characteristic impedance of the supply line can restrain the interference in the power line.

3.2. EMC design of computer system

*3.2.1. EMC design of I/O in computer.*The I/O unit is responsible for driving, level conversion, and ESD protection. We should avoid lacking the enough power of driving and ensure the electrostatic protection. The following focus on the ESD protection.

ESD protection can be obtained through components such as Transient Voltage Suppression (TVS) diodes, Polymer diodes and Varistors. Polymer diodes and Varistors offer huge residual pulse and hence are not preferred.

TVS diodes for a particular interface are selected based on many characteristics. Few of the key characteristics are stated below.

1. ESD protection level: IEC 61000-4-2 standard specifies level 2 for tablet products. TVS diodes are chosen to withstand these ESD voltages.

2. Bidirectional or Unidirectional ESD protection: TVS diodes are to be chosen based on protection required for bidirectional or unidirectional signals.

3. Clamping voltage: Clamping voltage of TVS diodes must be lower enough to avoid soft failures.

4. Residual pulse: It is a short duration high voltage peak seen after the TVS diode clamps the ESD pulse. Care should be taken while choosing TVS diodes to avoid residual pulse leading to crosstalk and soft failure issues.

5. TVS Diode Capacitance: For high speed interfaces, TVS diodes which offer lower capacitance are preferred to allow for good signal integrity[4].

3.2.2. EMC design of switching power supply. With the increase of the switching frequency, the EMI problems become severer. They interfere not only to the environment and the equipment around them, but also to themselves. In order to improve the stability and reliability of the devices, the electro-magnetic compatibility (EMC) of the switch power supply is got more attention than ever before [5].

The peak voltage interference generated by voltage conversion circuit and current harmonics generated by the rectifier circuit to produce are the main factors that switching power supply produces electromagnetic harassment. Additionally, formation of switch pulse and control circuit in high frequency rectangular pulse will produce electromagnetic disturbance.

There are two measures of suppressing electromagnetic interference produced by switching power supply:

Firstly, reduce the harassment intensity of the disturbance source

(1) Select the appropriate switching power supply mode and frequency

In practical design, when selecting the work mode, we should consider the circuit complexity, the conversion efficiency, the production cost and the difficulty of debugging and its electromagnetic compatibility, which can often get an idea effect. As for the frequency of work, it should be smaller if not increases the cost and affects the efficiency of the work.

(2) Select the appropriate circuit components

The switching loop is the direct and major source of electromagnetic interference generated by switching power supplies. In the switching loop, the switch tube is the core. The design of the switch tube takes the high frequency suppression and the momentary shock of the switch into account. Another key component in the switching circuit is a pulse transformer. The influence of pulse transformer on electromagnetic compatibility is shown in two aspects: one is the distributed capacitance of primary coil and secondary coil, and the other is the leakage of pulse transformer. By adding the electrostatic shielding layer in the primary and secondary coils and leading it to ground, we can greatly reduce the switch capacitance between the primary and the secondary transformer, thereby reducing the coupling of the primary and secondary electric field disturbance.

Secondly, cut off harassment routes of transmission

(1) Using filtering technique

We can insert the common-mode and differential-mode filter in the switching power supply AC power line input end, which prevent the power line from common-mode and differential-mode noise.

*3.2.3. Chassis design.*The reasonable design of the chassis can restrain the electromagnetic interference and shield the electromagnetic wave outside the computer. Enabling the chassis to possess good continuity of conduction is the key of design.

The concrete method is to use the electromagnetic gasket with good conductivity fasten the joint of the shell board. Prevent the line from passing through the shield directly and use the cable socket with the proper filter.

4. EMC test

4.1 The basic content of EMC test

EMC testing is mainly composed of electromagnetic interference testing and electromagnetic sensitivity testing. When the tested intelligent controller in normal operation, it will produce electromagnetic wave signal, and electromagnetic interference testing is detecting electromagnetic signals emitted by intelligent controller [6].

4.2 EMC test method

Use the EMC radiation intensity tester to judge whether the computer system meets the regulatory requirements, namely the design in accordance with the "International Electro technical Commission" standards, including FCCRART15, EN55022, and GB9254 etc. In general, according to different types of products and where the products to sell the region, we can determine which standard to use. Then according to the standard, we can determine the corresponding test items.

5. Conclusion

With the spread of the computers, electromagnetic interference has become a problem that has to be solved. Through the analysis of electromagnetic compatibility of computer board and system, we can effectively take effective measures to avoid electromagnetic interference in the design stage, which means that we can improve the validity of EMC design and reduce the cost during design stage.

References

[1] Dan Dinu, Bernhard Auer. Design considerations for improved electromagnetic compatibility (EMC) behaviour in high side switches design[C]. Sinaia, Romania: Semiconductor Conference, 2009. CAS 2009. International, 2009. 417-420

[2] Mosin Mondal, Samuel Connor, Bruce Archambeault, Vikram Jandhyala. Fast frequency domain crosstalk analysis for board-level EMC rule checking and optimization [C]. Detroit, MI, USA: 2008 IEEE International Symposium on Electromagnetic Compatibility, 2008. 1-6

[3] Deepak Sharma, Geetha Vanga. PCB design guidelines of Smartphones and Tablets for EMC [C]. Bangalore, India: 2016 International Conference on ElectroMagnetic Interference & Compatibility, 2016. 1-3

[4] Geetha Vanga, Yagnesh V Waghela. ESD design methodology for tablets [C]. Bangalore, India: 2016 International Conference on ElectroMagnetic Interference & Compatibility, 2016. 1-3

[5] Chen Wei, Zou Yun-pin, Xu Li-juan, Lu Li. EMC Design of 10kW DC Switch Power Supply [C]. Dalian, China: The 2006 4th Asia-Pacific Conference on Environmental Electromagnetics, 2006. 181-184

[6] Shu Yang, Chen Li-an, Gao Xin-gen, Tian Hong;, Liu, Yan. Intelligent pole mounted switch controller EMC design and test [C]. Hangzhou, China: 2015 7th Asia-Pacific Conference on Environmental Electromagnetics, 2015. 214-220

IOP Conf. Series: Journal of Physics: Conf. Series **976** (2018) 012013 doi:10.1088/1742-6596/976/1/012013

A Horizontal Tilt Correction Method for Ship License Numbers Recognition

Baolong Liu[1,2], Sanyuan Zhang[1], Zhenjie Hong[2] and Xiuzi Ye[2,*]

[1]College of Computer Science and Technology, Zhejiang University, Hangzhou 310027, China

[2]College of Mathematics, Physics and Electronic Information Engineering, Wenzhou University, Wenzhou 325035, China

[*]Corresponding author (Yexiuzi@wzu.edu.cn)

Abstract. An automatic ship license numbers (SLNs) recognition system plays a significant role in intelligent waterway transportation systems since it can be used to identify ships by recognizing the characters in SLNs. Tilt occurs frequently in many SLNs because the monitors and the ships usually have great vertical or horizontal angles, which decreases the accuracy and robustness of a SLNs recognition system significantly. In this paper, we present a horizontal tilt correction method for SLNs. For an input tilt SLN image, the proposed method accomplishes the correction task through three main steps. First, a MSER-based characters' center-points computation algorithm is designed to compute the accurate center-points of the characters contained in the input SLN image. Second, a $L_1 - L_2$ distance-based straight line is fitted to the computed center-points using M-estimator algorithm. The tilt angle is estimated at this stage. Finally, based on the computed tilt angle, an affine transformation rotation is conducted to rotate and to correct the input SLN horizontally. At last, the proposed method is tested on 200 tilt SLN images, the proposed method is proved to be effective with a tilt correction rate of 80.5%.

1. Introduction

An SLN is composed by a sequence of characters. An automatic ship license numbers (SLNs) recognition system plays a significant role in intelligent waterway transportation systems since it can be used to identify ships by recognizing the characters in SLNs [1-2].

SLNs recognition is a brand-new research topic and few related papers have been published. [1] used a transferred deep CNN model in combination with prior features to localize multi-style ship license numbers in nature scenes. Liu et al. [2] presented an effective coarse-to-fine approach for locating various SLNs in the wild.

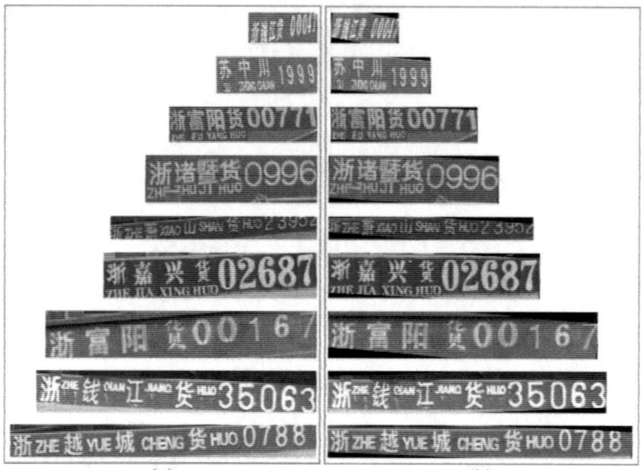

(a) (b)

Figure 1. Some tilt correction results of the proposed method. (a) Some SLN images in our testing dataset. (b) The corresponding tilt correction results.

One related research area of SLNs recognition is automatic license plate recognition (ALPR) [3]. Similar with an ALPR system, an automatic SLNs recognition system mainly consists of image acquisition, SLNs location, tilt correction, character segmentation and character recognition. Ships usually appear in the complicated environments of rivers and canals, while the position-fixed monitors mostly set on the banks. Under this circumstance, most SLNs in the captured images have serious tilt because the monitors on the bank and the ships on the water have a great vertical or horizontal angle [4]. These tilt SLNs will affect the accuracy and robustness of the following character segmentation and character recognition procedures of a SLNs recognition system significantly. Hence, an effective tilt correction method improves the performance of an SLN recognition system.

Tilt correction can be generally categorized into horizontal tilt correction and vertical tilt correction [5]. In this paper, we focus on horizontal tilt correction, and a horizontal tilt correction method of SLNs is presented. For an input tilt SLN, a MSER-based [6] characters center-points computation algorithm is firstly designed to compute the accurate center-points of the characters contained in the input SLN image. Then, a $L_1 - L_2$ distance-based straight line is fitted to these computed center-points using M-estimator algorithm to estimate the tilt angle of the input SLN. Finally, based on the computed tilt angle, an affine transformation rotation is conducted to rotate and to correct the input SLN horizontally.

The proposed method is experimented on 200 tilt SLNs which are collected from the SLNs dataset of ZJUSHIPS950 [2]. The experiment results in Sec. 4 indicate that the proposed approach achieves a horizontal tilt correction rate of 80.5%. The main contribution of this paper is that we present an effective horizontal tilt correction method for SLNs recognition, which, as far as we know, is the first work concerning the SLNs tilt correction problem.

2. Related works

Works that related most with SLNs tilt correction is car license plates tilt correction [4-5, 7-9]. Car license plates tilt correction methods can be roughly classified as detecting line methods [4], orientation image methods [7] and linear fitting algorithms [8]. Detecting line-based methods [4] correct tilt license plates by localizing the four sides of the plates using straight line detection algorithms, e.g., Hough transformation. An important prior of detecting line methods is the characters are always written on plates. However, detecting line methods are not applicable for SLNs tilt correction because the SLNs are usually written directly on the ship bodies rather than on plates. As a kind of orientation image-based tilt correction method, [7] used the local maximum of the direction angle histogram to estimate the horizontal incline angle of a license plate. It is also difficult for orientation image-based car license plates tilt correction methods to correct SLNs because characters contained in one single SLN may have different tilt angles. [8] is a linear fitting-based algorithm, it computes a horizontal rotation angle by fitting a straight line to the license plate using least square fitting with perpendicular offsets. Linear fitting-based algorithms usually assume that characters from one same car license plate are written in one line. However, the layouts of SLNs from different ships are complicated (Fig. 1(a)). Characters contained in one same SLN may be written in two lines, causing most linear fitting-based algorithms to obtain unreasonable straight lines. Our approach is also linear fitting-based, nevertheless, an extremal regions filtering and merging algorithm is posed in Sec. 3 to reduce the influences of two-line writing and noises. In conclusion, it is complicated to directly use car license plate tilt correction algorithms to perform SLNs tilt correction. To accurately correct tilt SLNs, a tilt correction method taking the characteristics of SLNs into consideration needs to be proposed.

3. The proposed method

Fig. 2 demonstrates the detail processing procedures of the proposed method. It accomplishes the horizontal tilt correction task through three main procedures: 1) MSER-based characters center-points computation. The extremal regions (ERs) of the input SLN image are firstly computed. Then, these ERs are filtered by referring their widths, heights and area sizes. Third, the reserved ERs are merged according to their horizontal distances, the center-points of the remaining ERs are obtained; 2) $L_1 - L_2$ distance-based straight line fitting. An $L_1 - L_2$ distance-based straight line is fitted to the computed center-points by using M-estimator algorithm. The tilt angle can be also computed at this stage; 3) Horizontal tilt correction. Based on the computed tilt angle, an affine transformation rotation is proceeded to rotate and correct the input SLN horizontally.

3.1 MSER-based characters center-points computation

The SLN image in Fig. 2 is a Chinese cargo ship SLN, some characters of that SLN are written in one line while others are in two lines. Actually, the English characters below the Chinese characters are phonetic transcriptions, they are not required to be recognition. Moreover, the existence of those smaller phonetic transcription-characters may decrease the accuracy of the straight lines fitted to the characters need to be recognized. To improve the center-points computation accuracy, it is necessary to remove these small phonetic transcription-characters and other non-character regions.

The details of our MSER-based characters center-points computation algorithm are summarized as follows:

(1) MSER-based character regions computing. The MSER [6] method is primarily used to compute the ERs of each input SLN image.

(2) ERs filtering. The obtained ERs are then filtered by referring the geometrical characteristics of characters. Specifically, for each obtained ER, an axis-oriented minimum bounding box is estimated, the ratios between the width, height and area size of the estimated bounding box and the width, height and area size of the corresponding input SLN image are computed, respectively. If one of these computed ratios is too small or too great to a certain range, this ER will be filtered.

(3) ERs merging and center-points computing. In this step, the intersected ERs are horizontally merged. All bounding boxes are sorted by the values of their Coordinate X in ascending order. Then, if a bounding box and its previous adjacent bounding box are intersected with each other and the area size of the intersected region is greater than one-third the area size of the previous bounding box, these two bounding boxes will be merged as a new one. At last, the center-points of the remaining bounding boxes are computed.

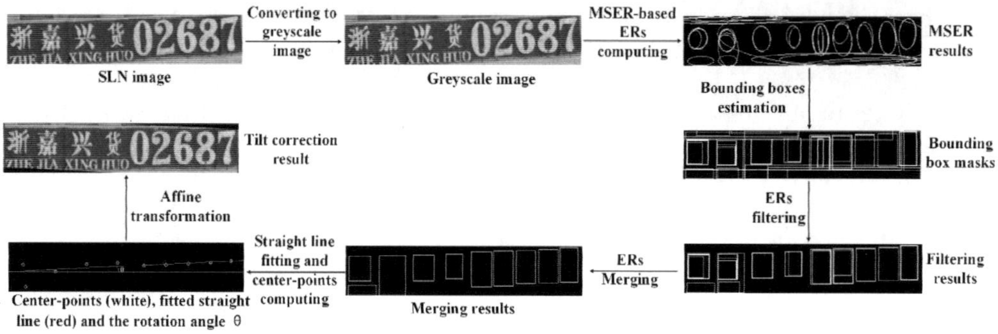

Figure 2. The detail processing procedures of the proposed method.

3.2 $L_1 - L_2$ distance-based straight line fitting

In order to compute the rotation angle, a straight line is fitted to the center-points computed in Sec. 3.1 by referring [2]. Assume that $P_c = \{(x_1, y_1), (x_2, y_2), \ldots, (x_i, y_i)\}$ is the set of the center-points computed from an SLN image. A line $f(x)$ is fitted to P_c using the M-estimator algorithm [10], where $f(x)$ needs to meet the condition of minimizing $\sum_i \rho(r_i)$. r_i is the distance between the ith point in P_c and the fitting line $f(x)$. $\rho(r)$ is the $L_1 - L_2$ distance function of the M-estimator, which is defined as

$$\rho(r) = 2 \cdot \left(\sqrt{1 + \frac{r^2}{2}} - 1\right). \tag{1}$$

Finally, the slope of the fitted straight line $f(x)$ is used as the rotation angle.

3.3 Affine transformation-based horizontal tilt correction

Since the rotation angle has already been computed in the previous step, the input SLN image is rotated by conducting an affine transformation by referring the estimated rotation angle. The affine transformation function in OpenCV (https://opencv.org/) is used here.

4. Experimental results and analysis

4.1 Testing dataset

The effectiveness of the proposed approach was verified on 200 tilt SLN images. These tilt SLN images were captured from different ship images in ZJUSHIPS950 SLNs dataset [2]. These tilt SLN images are of the characteristics of complex backgrounds, random photographing angles, different illuminations and various layouts. Fig. 1 shows some testing SLN images.

4.2 Experiment results

The proposed method was implemented using C++ and was tested on a computer with an Intel 3.30 GHz CPU configured. An output SLN image is considered as a successfully corrected one if the characters contained in that SLN are observed to be in a fairly horizontal line, otherwise it will be treated as a failure. The quantitative experiment results are listed in Table 1. Table 1 tells that the proposed approach achieves a horizontal tilt correction rate of 80.5%, suggesting the proposed method has good tilt SLNs correcting ability. Table 1 also demonstrates that the proposed algorithm achieves a very fast time efficiency of 3.6 milliseconds per SLN image. One can also recognize the good tilt SLNs correcting performances of the proposed approach from Fig. 1. In addition, the tilt correction results in Fig. 3(a)-3(c) suggest that our method can still achieve fine results even when occlusion, blur or poor image quality occur.

Figure 3. Some sample tilt correction results of SLN images with occlusion (a), poor quality (b) and blur (c). (d) and (e) demonstrate some failure correction results.

Table 1. Horizontal tilt correction results of SLNs

Number of SLN images	Number of valid corrections	Number of failures	Tilt correction rate	Time (ms/SLN image)
200	161	39	80.5%	3.6

4.3 Discussion and failures

Our approach computes rotation angles by fitting a straight line to all detected characters regions, thus, if a few characters are occluded, the remaining characters could provide clues to fit a fairly reasonable straight line. Fig. 3(a) gives some correction results of SLNs with occlusion, showing that our approach has good ability to correct SLN images with occlusion.

Blur weakens the edge features of characters, causing the MSER method to compute many overlapping ERs. Fortunately, the ERs filtering algorithm in Sec. 3 is helpful to alleviate this. After ERs filtering, a more accurate straight line can be obtained since the redundant and interrupted ERs are removed. From Fig. 3(c) one can conclude that our method is capable of accurately correcting SLNs images with blur in many cases. However, as shown in Fig. 3(d), failures may appear when serious blur occurred.

The experimental results in Fig. 3(b) show that our approach is fairly robust to poor-quality images, it can correct tilt SLNs even when some strokes of the characters are missing or interfered. Notwithstanding, if the resolutions of SLN images are extremely poor or too much noise information existed, the proposed approach may fail to correct SLNs (Fig. 3(e)). The reason is that it is hard for MSER algorithm to detect poor-resolution scripts as maximally stable extremal regions.

5. Conclusion

We have proposed an effective horizontal tilt correction method of SLNs in this paper. The proposed method was tested on 200 tilt SLNs images from ZJUSHIPS950 dataset, as the first work concerning the SLNs tilt correction problem, the proposed method is proved to be effective with a tilt correction rate of 80.5%. More SLNs prior features will be exploited in the future work to accurately correct tilt SLNs.

6. Acknowledgments

This work was supported partially by the National Natural Science Foundation of China under Grant no. 61272304 and partially by the Zhejiang Provincial Natural Science Foundation of China under Grant no. LY16F020023.

IOP Conf. Series: Journal of Physics: Conf. Series **976** (2018) 012013 doi:10.1088/1742-6596/976/1/012013

References

[1] B. Liu, X. Lyu, C. Li, S. Zhang, Z. Hong, and X. Ye, "Using transferred deep model in combination with prior features to localize multistyle ship license numbers in nature scenes," in *IEEE International Conference on TOOLS with Artificial Intelligence*, 2018, pp. 506–510.

[2] B. Liu, J. Sheng, J. Dun, S. Zhang, Z. Hong, and X. Ye, "Locating various ship license numbers in the wild: An effective approach," *IEEE Intelligent Transportation Systems Magazine*, vol. 9, no. 4, pp.102–117, 2017.

[3] S. Du, M. Ibrahim, M. Shehata, and W. Badawy, "Automatic license plate recognition (alpr): A state-of-the-art review," *IEEE Transactions on Circuits and Systems for Video Technology*, vol. 23, no. 2, pp. 311–325, 2013.

[4] S. J. Yang, C. C. Ho, J. Y. Chen, and C. Y. Chang, "Practical homography-based perspective correction method for license plate recognition," in *International Conference on Information Security and Intelligence Control*, 2013, pp. 198–201.

[5] Z. Fu, Y. Zhao, X. Li, and W. Guo, "Character string envelope correlation based vehicle license plate tilt correction in vertical direction," in *IEEE Information Technology, Networking, Electronic and Automation Control Conference*, 2016, pp. 160–164.

[6] J. Matas, O. Chum, M. Urban, and T. Pajdla, "Robust wide-baseline stereo from maximally stable extremal regions," *Image and Vision Computing*, vol. 22, no. 10, pp. 761–767, 2004.

[7] R. Pan, X. Ma, and L. Wang, "An efficient method for skew correction of license plate," in *Second International Workshop on Education Technology and Computer Science*, 2010, pp. 90–93.

[8] K. Deb, A. Vavilin, J. W. Kim, T. Kim, and K. H. Jo, "Projection and least square fitting with perpendicular offsets based vehicle license plate tilt correction," in *Sice Conference 2010, Proceedings of*, 2010, pp. 3291–3298.

[9] M. Pan, J. Yan, and Z. Xiao, "An approach to tilt correction of vehicle license plate," in *International Conference on Mechatronics and Automation*, 2007, pp. 271–275.

[10] V. P. Godambe and B. K. Kale, "Estimating functions: an overview," 1991.

A Sector Capacity Assessment Method Based on Airspace Utilization Efficiency

Jianping Zhang[1], Ping Zhang[2], Zhen Li[3] and Xiang Zou[1]

[1] The Second Research Institute of Civil Aviation Administration of China, Chengdu, Sichuan, China, 610041
[2] Guizhou Air Traffic Management Sub-bureau of Civil Aviation of China, Guiyang, Guizhou, China, 550012
[3] Chongqing Air Traffic Management Sub-bureau of Civil Aviation of China, Chongqing, China, 401120

E-mail of the corresponding author: zouxiang@caacsri.com

Abstract. Sector capacity is one of the core factors affecting the safety and the efficiency of the air traffic system. Most of previous sector capacity assessment methods only considered the air traffic controller's (ATCO's) workload. These methods are not only limited which only concern about the safety, but also not accurate enough. In this paper, we employ the integrated quantitative index system proposed in one of our previous literatures. We use the principal component analysis (PCA) to find out the principal indicators among the indicators so as to calculate the airspace utilization efficiency. In addition, we use a series of fitting functions to test and define the correlation between the dense of air traffic flow and the airspace utilization efficiency. The sector capacity is then decided as the value of the dense of air traffic flow corresponding to the maximum airspace utilization efficiency. We also use the same series of fitting functions to test the correlation between the dese of air traffic flow and the ATCOs' workload. We examine our method with a large amount of empirical operating data of Chengdu Controlling Center and obtain a reliable sector capacity value. Experiment results also show superiority of our method against those only consider the ATCO's workload in terms of better correlation between the airspace utilization efficiency and the dense of air traffic flow.

1. Introduction

Air traffic flow management (ATFM) is a service established with the objective of contributing to a safe, orderly and expeditious flow of air traffic by ensuring that air traffic control (ATC) capacity is utilized to the maximum extent possible [1]. Therefore, ATC sector capacity assessment should be the precondition and basis for implementing ATFM.

In last decades, studies on ATC sector capacity assessment have been mostly dedicated to examine the relationship between the capacity and the workload of air traffic controllers [2-5]. Although ATCO's workload can be used to evaluate ATC sector capacity in terms of either the number of simultaneously present aircraft or the number of aircraft traversing the sector per unit of time, the sector capacity is also affected by some other factors. Ball et al. pointed out that the capacity of a sector presents variability and unpredictability. He summarized the factors into three categories, which were airspace physical structure, air traffic situation, and operational constraints [6]. In fact, estimating

IOP Conf. Series: Journal of Physics: Conf. Series **976** (2018) 012014 doi:10.1088/1742-6596/976/1/012014

the sector capacity by ATCO's workload mainly concerns about the safety. However, as stated in the first paragraph, the sector capacity has significant impact on the ATFM, which aims at utilizing the airspace to the maximum extent. Hence, from the ATFM point of view, the sector capacity should reveal how efficiently the airspace is utilized. In addition, it may not be accurate if we estimate the capacity only by ATCO's workload. In this paper, we try to solve the problems mentioned here and improve the accuracy, reliability, and applicability of capacity estimation methods through the points as follows.

(1) We employ the integrated index system proposed in one of our previous literatures [7], which covers 14 indicators about operational performance, airspace complexity handling, safety, economic efficiency, and ATCO's. We use the PCA to analyze the indicators in order to dig out the most important ones that have impact on the airspace utilization efficiency and to calculate the score of the airspace utilization efficiency.

(2) We use a fitting function to define the correlation between the airspace utilization efficiency and the dense of air traffic flow which is defined as the quantity of flights traversing a sector during a unit of time. We set the sector capacity as the value of the dense of air traffic flow corresponding to the maximum airspace utilization efficiency. We also show better correlation between the airspace utilization efficiency and the dense of the flow than that between the ATCO's workload and the dense of the flow by some statistical attributes of the fitting functions.

2. Multivariate Index System for airspace utilization efficiency

Airspace utilization efficiency of an ATC sector should reveal its relationship with both the safety and the operating efficiency. We put forward an improved integrated and universal multivariate index system systematically for the airspace utilization efficiency of an ATC sector based in one of our previous literatures [7]. The notations and the definitions of the indicators are listed in table 1.

Table 1. Multivariate index system for Airspace utilization efficiency of an ATC sector

Variable	Indicator	Definition	Notation (Unit)
Performance Efficiency	Flight Miles	Total flight miles in sector per hour	X_1 (km)
	Flight Time	Total flight time in sector per hour	X_2 (min)
Complexity Handling Efficiency	Aircraft Climbing Frequency	Total aircraft climbing times in sector per hour	X_3 (time)
	Aircraft Descending Frequency	Total aircraft descending times in sector per hour	X_4 (time)
	Aircraft Speed Changing Frequency	Total aircraft speed changing times in sector per hour	X_5 (time)
	Aircraft Heading Changing Frequency	Total aircraft heading changing times in sector per hour	X_6 (time)
Safety Efficiency	STCA Frequency	Total STCA times in sector per hour	X_7 (time)
	MSAW Frequency	Total MSAW times in sector per hour	X_8 (time)
Economic Efficiency	Sector Queue Length	Total waiting flight number in sector per hour	X_9 (flight)
	Delay Percentage	Ratio of delayed flight number and total flight number in sector per hour	X_{10} (%)
	Delay Time	Total delay time of flights in sector per hour	X_{11} (min)
	Average Delay Time	Average delay time of total delayed flights in sector per hour	X_{12} (min)

ATCO Workload	occupation percentage of air-ground communication radio	Occupation percentage of air-ground communication radio between ATCO and pilots in sector per hour	X_{13} (%)
	call times of air-ground communication radio	Call times of air-ground communication radio between ATCO and pilots in sector per hour	X_{14} (time)

3. The processes of sector capacity assessment

We set 3 basic steps to estimate the capacity of an ATC sector as follows.

Step 1: Set the data set.

Collect operating data of an ATC sector hourly and calculate the values of those 14 indicators in Table 1. Each sample data represents the state of an ATC sector during single hour.

Step 2: Principal components analysis.

Based on the sample data, apply principal component analysis to reduce the dimension of the data and calculate the score of the airspace utilization efficiency of the sector.

Step 3: Function fitting.

Fit the dense of the air traffic to the evaluated score of the airspace utilization efficiency. Set the sector capacity as the value of the dense of the air traffic flow corresponding to the highest value of the airspace utilization efficiency.

4. The PCA and the function fitting

4.1. Principal component analysis

We use the principal component analysis (Amos G, 2008)18 to reduce the dimension of the collected data set and calculate the value of the airspace utilization efficiency. The procedures are as follows:

(1) The preprocessing of the source data.

The input data should be normalized to eliminate the dimension differences among all indicators. Set x_{ij} as the value of indicator j ($j = 1, 2, ..., 14$) of i_{th} ($i = 1, 2, ..., N$) sample data.

The empirical mean of indicator j is

$$\overline{X_j} = \frac{1}{N}\sum_{i=1}^{N} x_{ij} \tag{1}$$

The standard deviation of indicator j is

$$\sqrt{V(X_j)} = \sqrt{\frac{1}{N-1}\sum_{i=1}^{N}(x_{ij} - \overline{X_j})^2} \tag{2}$$

Then the normalized matrix, $Z = (z_{ij})_{N\times 14}$, which is computed using the empirical mean and standard deviation of each indicator is composed with elements like

$$z_{ij} = \frac{x_{ij} - \overline{X_j}}{\sqrt{Var(X_j)}} \tag{3}$$

(2) Compose the correlation matrix.

Set $R = (r_{j,j'})_{14\times 14}$ as the correlation matrix. Its element, $r_{j,j'}$ denotes the correlation between the j_{th} indicator and the j'_{th} indicator, where

$$r_{j,j'} = \frac{\sum_{k=1}^{N} z_{kj} z_{kj'}}{N-1} = \frac{\sum_{k=1}^{N} (x_{kj} - \overline{X_j})(x_{kj'} - \overline{X_{j'}})}{\sqrt{\sum_{k=1}^{N} (x_{kj} - \overline{X_j})^2 \sum_{k=1}^{N} (x_{kj'} - \overline{X_{j'}})^2}} \quad j, j' = 1, 2, \dots, 14 \tag{4}$$

(3) Calculate the eigenvectors and eigenvalues of the correlation matrix.

Set the eigenvalues of correlation matrix as $\lambda_1 \geq \lambda_2 \geq \dots \geq \lambda_{14}$ with corresponding eigenvectors array $U = (U_1, U_2, \dots U_{14})^T$

(4) Calculate the evaluated scores.

Calculate the principal component value matrix as below,

$$Y = (y_{ij})_{N \times 14} = (Y_1, Y_2, \dots, Y_{14}) = Z * U \tag{5}$$

Calculate the contribute rate of the j_{th} eigenvalue as below,

$$w_j = \frac{\lambda_j}{\sum_{j=1}^{P} \lambda_j}, \ j = 1, 2, \dots, P \tag{6}$$

Choose m eigenvalues from the biggest to the smallest whose cumulative contribute rate is no less than 0,8,

$$\sum_{j=1}^{m} w_j \geq 0.8 \tag{7}$$

Calculate the evaluated scores of the air space utilization efficiency of each sample as,:

$$E = \sum_{j=1}^{m} w_j Y_j \tag{8}$$

4.2. Function fitting

We measure the dense of the air traffic flow with the quantity of flights traversing a sector during single hour. Then, we fit the values of the dense of the air traffic flow relative to the samples data to the evaluated scores of the airspace utilization efficiency with three different fitting functions: linear, polynomial, and Fourier function. We choose the function with least R-square or SSE (Sum Squared Error) values to define the correlation.

5. Empirical Study

Since Chengdu area is one of the busiest air traffic area in China, one sector in Chengdu area is taken as the typical example for empirical study.

5.1. Data set

We collect 648 sets of source data, each for single hour, during rush hours. Table 2 is an example of one sample data.

Table 2. A sample of the source data

Indicator	X_1	X_2	X_3	X_4	X_5	X_6	X_7
data	1679	334	5	13	55	158	119
Indicator	X_8	X_9	X_{10}	X_{11}	X_{12}	X_{13}	X_{14}
data	72	1	7.69%	12	12	42%	66

IOP Conf. Series: Journal of Physics: Conf. Series **976** (2018) 012014 doi:10.1088/1742-6596/976/1/012014

We apply MATLAB2014a to execute the principal component analysis and decompose 14 indicators into 6 principal components with the cumulative contribute rate of 0.8294.

5.2. Results of the function fitting

Figure1 to figure 3 shows the fitting images and parameters for three functions.

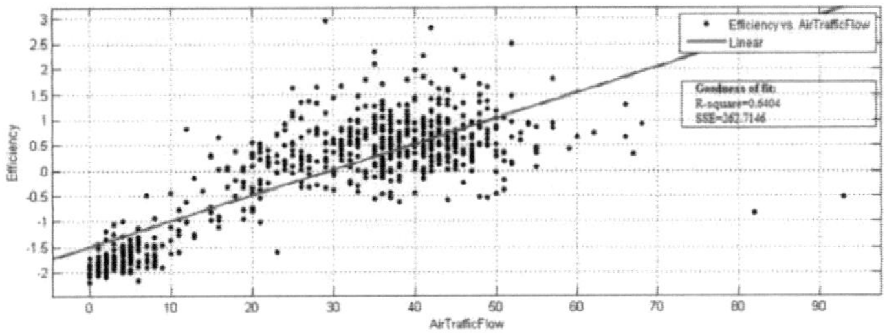

Figure 1 Fitting image by Linear function for the airspace utilization efficiency and the dense of the air traffic flow

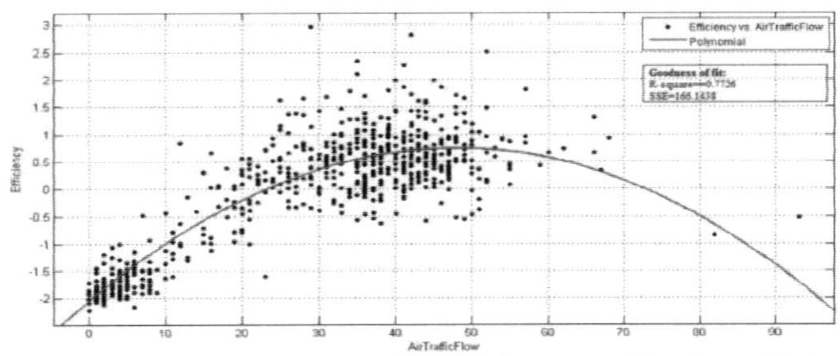

Figure 2 Fitting image by Polynomial function for the airspace utilization efficiency and the dense of the air traffic flow

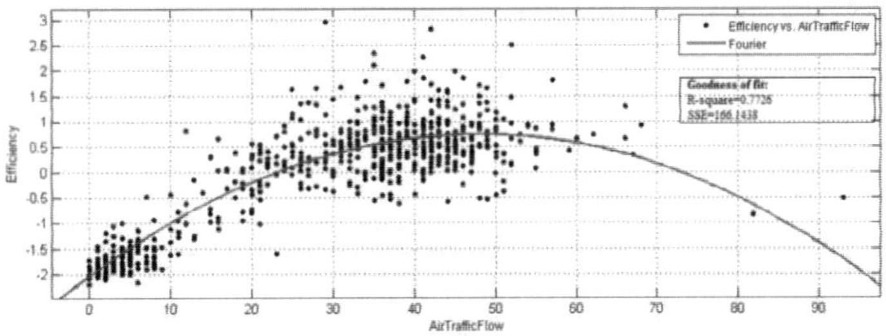

Figure 3 Fitting image by Fourier function for the airspace utilization efficiency and the dense of the air traffic flow

It's clear that the polynomial function and the Fourier function work better than the linear function in terms of higher R-square and lower SSE (Sum Squared Error). In detail, the R-square and the SSE of polynomial and Fourier functions are both 20.64% higher and 36.76% lower than those of the linear function. The R-square values for polynomial and Fourier fitting are both 0.7726, which reveals strong correlation between the airspace utilization efficiency and the dense of the air traffic flow. Since the fitting qualities of the polynomial function and the Fourier function are the same in terms of the values of the R-square and the SSE, we formulate both the functions below,

Polynomial function:

$$y = -0.001213x^2 + 0.1164x - 2.046 \tag{9}$$

Fourier function:.

$$y = -3.786 * 10^7 + 3.786 * 10^7 \cos(-8.006 * 10^{-6}x) - 1.454 * 10^4 \sin(-8.006 * 10^{-6}x) \tag{10}$$

where x represents the dense of the air traffic flow and y represents the score of the airspace utilization efficiency.

We derive both the functions to get the extreme points corresponding to the highest score of the airspace utilization efficiency. The value of the x - axis is 48 at the extreme point which means that the sector capacity for this case can be set as 48 flights per hour. We test the values of ATCOs' workloads corresponding to the value of 48 flights per hour. Results show that the workloads in such a flow dense are all under the safety threshold, which support the claim that our method can take care of both the safety and the efficiency.

Furthermore, the values of the R-square and the SSE for all 3 functions are pretty worse. R-square and SSE for both the polynomial function and the Fourier function are 0.4824 and 99300 and the results for the linear function are 0.4007 and 114980. This result obviously shows that the correlation between the ATCO's workload and the dense of the air traffic flow is worse that the correlation proposed in this paper, which supports the claim that the method proposed here can estimate the sector capacity more accurately than those only considered the ATCO's workload.

6. Conclusion

This paper proposed a sector capacity estimation method based on airspace utilization efficiency. We employ an integrated index system with 14 indicators in total, which helps the comprehensive evaluation of the airspace utilization efficiency. We make use of the principal component analysis to dig out the most important components among the 14 indicators and calculate the score of the airspace utilization efficiency. We employ a series of fitting functions to build up the correlation between the score (or value) of the airspace utilization efficiency and the dense of the air traffic flow. We derive the most fitting function to obtain its extreme point which corresponds to the highest airspace utilization efficiency. Hence, the sector capacity can be set as the value of the dense of the air traffic flow corresponding to the extreme point. We validated our method through empirical study based on real operating data of one sector in Chengdu area, China. We get a reliable value of the sector capacity and the experiment result also shows that the airspace utilization efficiency proposed here fits much better with the dense of the air traffic flow than traditional ATCO's workload. These results show the superiority of our method in terms of better accuracy, reliability, and applicability in the field of sector capacity estimation.

References

[1] ICAO, Doc4444 (15th Edition)-2007, Procedures for Air Navigation Services-Air Traffic Management [S], Montreal, Canada: ICAO, 2007.
[2] Richmond G C. (1988). An interim description of the DORATASK methodology for Assessment of Sector Capacity. DORA Report 8818.London: Civil Aviation Authority.
[3] Richmond G C. (1989a). The DORATASK Methodology of Sector Capacity Assessment: an Interim Description of its Adaptation to Terminal Control (TMA) Sectors. DORA Report 8916. London: Civil Aviation Authority.
[4] ICAO. Doc9426-AN/924 (1st Edition)-1984. Air transport service planning manual, Part II, Appendix C [S].Montreal, Canada: ICAO, 1999.
[5] Pellegrini P, Rodriguez J. Single European sky and single European railway area: A system level analysis of air and rail transportation [J].Transportation Research – Part A: Policy and Practice, 2013, 57 (1): pp. 64–86.
[6] Ball M, Barnhart C, Nemhauser G, et al. Air transportation: irregular operations and control [J]. Handbooks in Operations Research and Management Science, 2007, 14(1): 1–67.
[7] Zhang J, Duan L, Guo J, et al. A Genetic Algorithm Based BP Neural Network Method for Operational Performance Assessment of ATC Sector [J]. Promet - Traffic & Transportation, 2016, 28(6):pp. 563–574.

IOP Conf. Series: Journal of Physics: Conf. Series **976** (2018) 012015 doi:10.1088/1742-6596/976/1/012015

HealthBand for Dementia Patients: Fall and Scream Detector and Caretaker Helper

Zeeshan Alam[1], Huma Samin[1] and Omar Bin Samin[1]
[1]Institute of Management Sciences, Peshawar, Pakistan

[1]omar.samin@imsciences.edu.pk

Abstract: The ratio of dementia patients is escalating with time and requires proper attention to help the people suffering from it to continue their activities of daily living (ADL). Such patients suffer from the symptoms like irregular sleep patterns, restlessness, wandering, screaming, falling, sadness and depression. Assistive Technology facilitates caretaker to aid the patient efficiently with minimum effort. Advances in technology have made possible state of the art and innovative methods of health care delivery. Home telecare; in which the patient's health is monitored remotely at home, is one such method. This paper is proposing a cost effective and user friendly wearable product based solution (i.e. HealthBand) that monitors patient's activities (specifically fall and scream) and notifies the caretaker in case of emergency to take appropriate action(s). These notifications are sent to the caretaker on the basis of predefined threshold and time span over Bluetooth and GSM mediums to android based application. The android app also keeps patient's medicines' intake record and reminds caretaker regarding medicine dosage and timings.

1. Introduction

Dementia especially Alzheimer's is a heart breaking disease that causes the loss of cognitive functions affecting the patient's daily life activities. Such patients suffer from the symptoms like irregular sleep patterns, restlessness, wandering, shouting, higher rates of falls, sadness and depression [1]. The higher fall rate is quite dangerous as it causes injuries in such patients thus adversely affects their health. People with dementia may shout, moan or use abusive language and their speech rate is also high. The caretaking of such patients becomes quite difficult for the caretakers and causes physical as well as emotional strain for them. The adoptions of assistive technology for detecting the fall and shout can help improve the caretaking process and quality of patient's life.

The use of wearable device (i.e. HealthBand) in order to monitor the day to day activities of a person for the health care is quite of importance. The use of technology like accelerometer plays a vital role in this regard. The use of wearable device (i.e. HealthBand) in order to monitor the day to day activities of a person for health care is quite of importance. The use of technology like accelerometer plays a vital role in this regard. Conventional fall detection systems propose the usage of MEMS accelerometers and tri-axial accelerometer in wireless adhoc networks in order to detect the fall of a person and transfer the fall information through the networked infrastructure; in the form of RF Signals; to the caretaker in order to take appropriate action[2].The tri-axial accelerometer; in waist mounted systems; is used to measure the parameters like patient's posture, energy expenditure and movement[3].In order to distinguish the actual fall from activities of daily living (ADL),some systems also propose usage of gyroscopes along with tri-axial accelerometers. For this purpose, human activities are divided into two categories of static postures like standing, bending, sitting, and lying and dynamic transitions are the movements and motions between the different static postures. The accelerometers are used to detect static postures and gyroscopes are used to detect dynamic transitions[4].A number of smart phone and smart watch basedfall detection systems;using built in sensors; have also been presented that monitor the patient's movements for recognizing fall [5][6].Some systems also apply machine learning algorithms in order to classify the two categories of fall in order to increase the accuracy of actual fall detection[7]. While others evaluate the data gathered from the accelerometer with threshold based algorithms.

Along with the position and fall detection sensors, the use of sound detection sensors plays a vital role in providing assistance to the patients. A number of systems; attached with the remote monitoring systems; dealing with speech analysis and classification of sounds have been presented. Such systems perform the task of speech recognition to distinguish a distress situation from a normal one[8].A number of systems present the idea of health smart home having microphones installed as an activity recognition system in order to detect environmental sounds[9, 10]. In order to detect scream, audio features like sound energy; specifically log energy and pitch are used. The system categorizes sound as a scream when there is continuity in it for a longer interval than a normal sound. [11]

A number of systems using ultra-sonic technology have also been presented. The ultra-sonic sensors are connected to devices like Arduino microcontroller or field programmable gate array (FPGA) processor in order to send the signal to the main processing unit using a networked architecture. The signal is analyzed by sensing the distance and detecting the actions like standing and falling etc. [12-14]

The main contribution of this paper is to present a unified solution "HealthBand" for fall and scream detection and medication and emergency alert. The product can monitor the patient's activities at home and unfamiliar surroundings as compared to the solutions that are limited to certain boundary. The product will also monitor occurrenceof hyper activity of the patient to check the effectiveness of the medication he is taking and also keeps medicine and activity record. This project also detects the fall duration to check the severe condition of the injury.

Section 2 of the paper discusses in detail the method and materials of the product that is being presented, Section 3 outlines the implementation of hardware, Section 4 provides an insight on hardware testing and finally the last section i.e. Section 5 concludes the paper.

2. Method and Materials

HealthBanduses the 3-axis Accelerometer and Microphone sensors to sense the fall and hyperactivity of the patient based on a threshold approach to detect falls and motionless periods. The following are the details of the materials used to develop theHealthBand.

2.1 Arduino Nano

Arduino is an open source computer hardware and software company for building intelligent electronic projects. It consists of physical programmable circuit board and an Integrated Development Environment (IDE), which is used to write and upload code from computer to the physical board.Wide range of entry level, enhanced features, Internet of Things(IoT) and educational Arduino boards are available in market. Board can be chosen according to specifications of each microcontroller. For this research, Arduino Nano is used because of its smaller size, optimum number of I/O pins and its compatibility with accelerometer. [15]

2.2 Accelerometer

ADXL-345 is a small and low power 3 axis accelerometer with high resolution measurement up to ±16g. Several special sensing functions are provided in ADXL-345. i.e. double and signal taps, activity and inactivity detection and free fall detection.In this research ADXL-345 is used to detect human fall. When the subject falls, the acceleration is rapidly changing and the angular velocity produces a variety of signals along fall direction. The lower and upper fall thresholds for these signals are used to identify the fall. [16]

IOP Conf. Series: Journal of Physics: Conf. Series **976** (2018) 012015 doi:10.1088/1742-6596/976/1/012015

2.3 Microphone Sensor

Microphone sensor module KY – 038 is used to detect patient's screaming. This sensor simply converts analog sound to electrical signal. Thresholds for these electrical signals are set to identify screaming. Once signals from microphone exceeds threshold, alarm is generated. [17]

2.4 Bluetooth Module

In this research, Bluetooth module is used for communication between HealthBand and Android application. Variety of Bluetooth serial interfaces are available such as; HC05 and HC06.Bluetooth module (HC05) is used as it can act as both master and slave whereas HC06 functions only as slave. [18]

2.5 GSM Module

GSM Module (900A) is used to provide long range communication between HealthBand and caretaker.

2.6 Android Application

The product contains a module of android application for remote monitoring of the patient by caretakers. The application receives signals from the HealthBand worn by dementia patient and notifies the caretaker in case of emergency. The application processes the signals for both fall and screams and generates alerts accordingly. In order to make the caretaking process easy, the application also contains the module for medicine intake reminder. With the help of medicine reminder module, one can maintain the pill dosage schedule and medicine intake history of the patient.

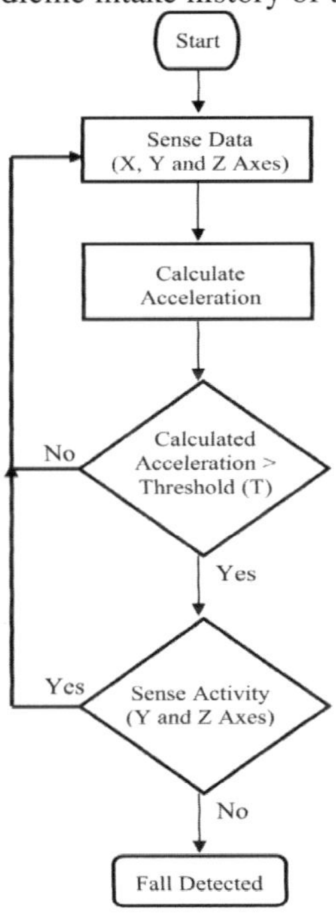

Figure 1: Fall Detection Data Flow Diagram

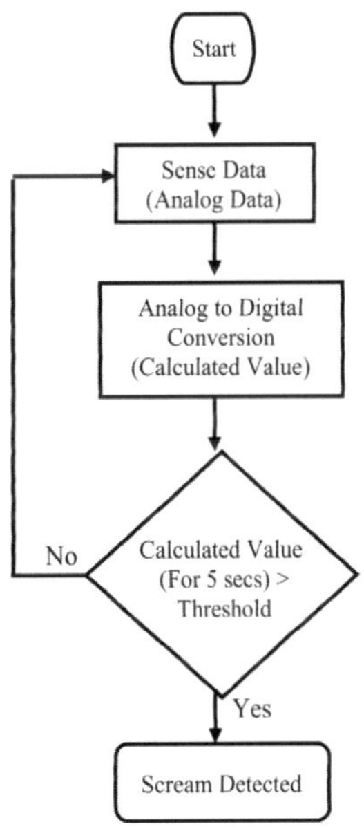

Figure 2: Scream Detection Data Flow Diagram

2.7 Fall Detection

Figure 1 explains the process of patient's fall detection. Accelerometer (ADXL345) detects movements along X, Y and Z axis (if any) and passes the data to microcontroller, where body acceleration is calculated. This calculated acceleration is compared with the pre-defined threshold, if this calculated value exceeds threshold, accelerometer further checks Y and Z axis for activity to reduce the possibility of false alarm. If activity is not sensed on Y and Z axis "fall" is detected.

IOP Conf. Series: Journal of Physics: Conf. Series **976** (2018) 012015 doi:10.1088/1742-6596/976/1/012015

2.8 Scream Detection

Figure 2 explains the process of patient's scream detection. Microphone sensor (KY-038) detects sound (if any) and passes the data to microcontroller, where analog to digital conversion of data takes place. This converted digital data (calculated value) is compared with the pre-defined threshold, if this calculated value exceeds threshold for 5 seconds "Scream" is detected.

3. Hardware Implementation

Figure 3 shows the connections amongst Arduino Nano, 3- Axis Accelerometer (ADXL-345), Microphone (KY-038), Bluetooth Module (HC-05) and GSM Module (900A).

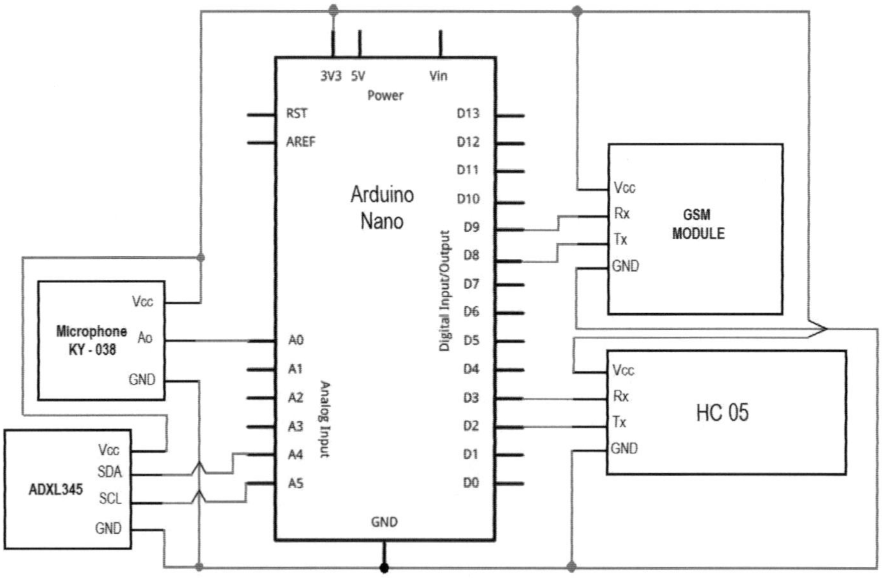

Figure 3: HealthBand Modules/ Components Connections

4. Hardware Testing

4.1 Fall Detection

Human body experience weightlessness during free fall and vector sum of acceleration on three axes tends towards 0gram.The duration of weightlessness depends on the height of freefall. Furthermore it's noteworthy that the weightlessness during a normal fall is not as significant as a freefall .i.e. vector sum of acceleration for normal fall is greater than 1 gram while for freefall is considerably less than 1 gram. Therefore, the first condition to detect freefall is weightlessness.When human subject experience weightlessness (Figure 4a)*(Blue: X Axis, Red: Y Axis and Grey: Z Axis)* and hits the ground; the vector sum of acceleration curve shows a large shock (Figure 4b)*(Blue: X Axis, Red: Y Axis and Grey: Z Axis).* This shock is identified by the ACTIVITY interrupt of ADXL345. Therefore, the second condition to detect fall is the ACTIVITY interrupt right after freefall.After fall and making impact to the ground human subject cannot rise immediately and remains motionless for short period of time. This motionlessness is identified by the INACTIVITY interrupt of ADXL345.Therefore, the third condition to detect fall is the INACTIVITY interrupt right ACTIVITY interrupt.Due to change in position after fall as compared to initial position accelerometer reading also changes. The acceleration data in all axes after the INACTIVITY interrupt is compared to the threshold, if vector sum of acceleration of the final status exceeds a certain threshold, an ACTIVITY interrupt occurs, since the static of the acceleration has changed after determining a fall to normal situation. Figure 4b and Figure 5*(Blue: X Axis, Red: Y Axis and Grey: Z Axis)* shows acceleration change curves during the fall of human subject.

IOP Conf. Series: Journal of Physics: Conf. Series **976** (2018) 012015 doi:10.1088/1742-6596/976/1/012015

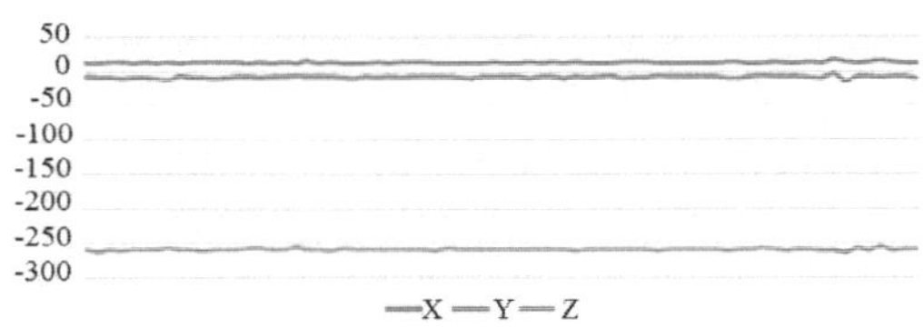

Figure 4a: Accelerometer Reading DuringWeightlessness

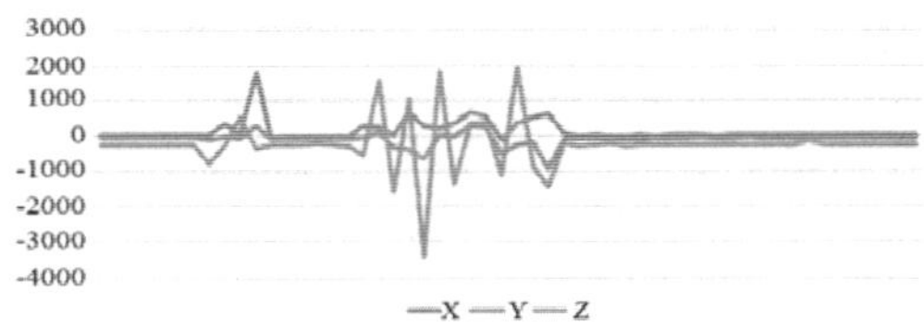

Figure 4b: Accelerometer Reading after Impact to the ground (ACTIVITY Interrupt)

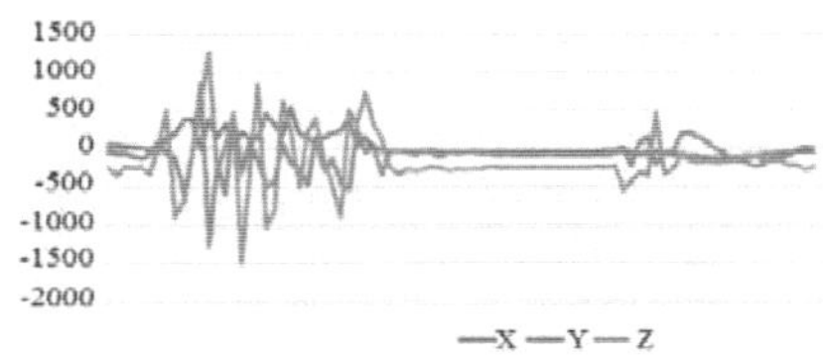

Figure 5: Accelerometer Reading before & after INACTIVITY Interrupt

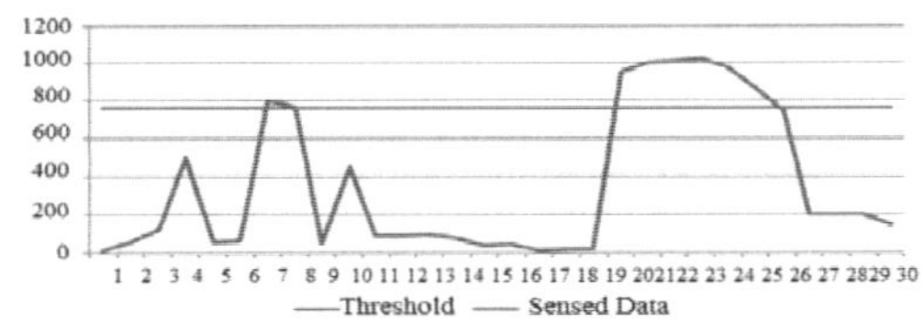

Figure 6*:* Scream Detection

4.2 Scream Detection

Occasionally (sometimes frequently) dementia patients scream, moan or use abusive language. The patient may become hyper due to pain, discomfort, under/ over stimulation, loneliness, anxiety and an unaccommodating environment (hot, cold or dark). Figure 6 (*Blue: threshold magnitude and Red: detected magnitude*) shows the frequency detected by microphone when patient suddenly gets hyper and starts screaming. If the value exceeds the threshold continuously for 5 seconds, scream is detected and alert signal is sent to caretaker on Android application to take required action.

5. Conclusion

The main focus of the paper is to present wearable product that is Hardware and Smartphone based solution to help the patients suffering from Dementia. The product monitors the fall and hyperactivity of the patients and alerts caretaker in case of emergency.The product also includes an android application to facilitate the caretaker to keep a check on patient's medicine intake, history and its effectiveness. It helps the patient/ caretaker in any environment (Indoor/ outdoor);not limiting to the surroundings of home only, with assistance. In future, we also want to include the fall direction detection and perform different analytics on the patient's medicine intake history as well as on the fall and hyperactivity durations in order to facilitate the doctors to check the effectiveness of the treatment and medicine intake.

References

[1] P. McCullagh, W. Carswell, J. Augusto, S. Martin, M. Mulvenna, H. Zheng, *et al.*, "State of the art on night-time care of people with dementia," 2009.

[2] J. Chen, K. Kwong, D. Chang, J. Luk, and R. Bajcsy, "Wearable sensors for reliable fall detection," in *Engineering in Medicine and Biology Society, 2005. IEEE-EMBS 2005. 27th Annual International Conference of the*, 2006, pp. 3551-3554.

[3] M. Mathie, J. Basilakis, and B. Celler, "A system for monitoring posture and physical activity using accelerometers," in *Engineering in Medicine and Biology Society, 2001. Proceedings of the 23rd annual international conference of the IEEE*, 2001, pp. 3654-3657.

[4] Q. Li, J. A. Stankovic, M. A. Hanson, A. T. Barth, J. Lach, and G. Zhou, "Accurate, fast fall detection using gyroscopes and accelerometer-derived posture information," in

Wearable and Implantable Body Sensor Networks, 2009. BSN 2009. Sixth International Workshop on, 2009, pp. 138-143.

[5] T. Dinh and M. Chew, "Application of a commodity smartphone for fall detection," in *Automation, Robotics and Applications (ICARA), 2015 6th International Conference on*, 2015, pp. 495-500.

[6] R. L. Almeida, A. A. Macedo, Í. L. de Araújo, P. A. Aguilar, and R. M. Andrade, "WatchAlert: Uma evolução do aplicativo fAlert para detecção de quedas em smartwatches."

[7] B. Aguiar, T. Rocha, J. Silva, and I. Sousa, "Accelerometer-based fall detection for smartphones," in *Medical Measurements and Applications (MeMeA), 2014 IEEE International Symposium on*, 2014, pp. 1-6.

[8] L. Abdoune and M. Fezari, "Everyday life sounds database: telemonitoring of elderly or disabled," *Journal of Intelligent Systems,* vol. 25, pp. 71-84, 2016.

[9] M. Vacher, A. Fleury, F. Portet, J.-F. Serignat, and N. Noury, "Complete sound and speech recognition system for health smart homes: application to the recognition of activities of daily living," ed: In-Tech, 2010.

[10] S. Tremblay, D. Fortin-Simard, E. Blackburn-Verreault, S. Gaboury, B. Bouchard, and A. Bouzouane, "Exploiting Environmental Sounds for Activity Recognition in Smart Homes," in *AAAI Workshop: Artificial Intelligence Applied to Assistive Technologies and Smart Environments*, 2015.

[11] W. Huang, T. K. Chiew, H. Li, T. S. Kok, and J. Biswas, "Scream detection for home applications," in *Industrial Electronics and Applications (ICIEA), 2010 the 5th IEEE Conference on*, 2010, pp. 2115-2120.

[12] C. Nadee and K. Chamnongthai, "Multi sensor system for automatic fall detection," in *Signal and Information Processing Association Annual Summit and Conference (APSIPA), 2015 Asia-Pacific*, 2015, pp. 930-933.

[13] C. Nadee and K. Chamnongthai, "Octagonal formation of ultrasonic array sensors for fall detection," in *Signal and Information Processing Association Annual Summit and Conference (APSIPA), 2016 Asia-Pacific*, 2016, pp. 1-4.

[14] Y. Huang and K. Newman, "Improve quality of care with remote activity and fall detection using ultrasonic sensors," in *Engineering in Medicine and Biology Society (EMBC), 2012 Annual International Conference of the IEEE*, 2012, pp. 5854-5857.

[15] "Arduino Nano", *Store.arduino.cc*, 2017. [Online]. Available: https://store.arduino.cc/usa/arduino-nano. [Accessed: 07- Nov- 2017].

[16] *Sparkfun.com*,2017. [Online].Available:https://www.sparkfun.com/datasheets/Sensors/Accelerometer/ADXL345.pdf. [Accessed: 07- Nov- 2017].

[17] "Microphone Sensor (KY-038) - Summerfuel Robotics", *Sites.google.com*, 2017. [Online]. Available: https://sites.google.com/site/summerfuelrobots/arduino-sensor-tutorials/microphones-with-arduino. [Accessed: 07- Nov- 2017].

[18] R. others, "HC-05 Bluetooth Module (Master/Slave) : rhydoLABZ INDIA", *rhydoLABZ.com*, 2017. [Online]. Available: http://www.rhydolabz.com/wireless-bluetooth-ble-c-130_132/hc05-bluetooth-module-masterslave-p-1169.html. [Accessed: 07- Nov- 2017].